Modelling of Flood Propagation Over Initially Dry Areas

Proceedings of the Specialty Conference
co-sponsored by
ASCE-CNR/GNDCI-ENEL spa

Approved for publication by the International Activities
Committee of the American Society of Civil Engineers

held in Milan, Italy
at ENEL-DSR-CRIS
29 June - 1 July 1994

Edited by Paolo Molinaro and Luigi Natale

Published by the
American Society of Civil Engineers
345 East 47th Street
New York, New York 10017-2398

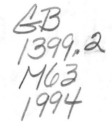

ABSTRACT

This proceedings, *Modelling of Flood Propagation Over Initially Dry Area,* contains papers presented at the Specialty conference co-sponsored by ASCE-CNR/GNDCI-ENEL spa and held in Milan, Italy, June 29 - July 1, 1994. The goal of the conference was to establish the state of the art in mathematical modelling of flood propagation over initially dry areas. The need for reliable evaluation tools of flood risk potential is paramount both in developed and developing countries. Developed areas need to study the reliability of their defense works while developing countries can use the mathematical models to plan their development in order to avoid future high risk situations. Although initially focused on the use of two-dimensional modelling of flood propagation, the subjects of the papers were broadened to include related topics such as geographical information systems. This proceedings presents papers covering 1) Theoretical and numerical aspects of flood modelling; 2) computational and practical aspects of modelling such as the use of GIS; 3) model verification against laboratory and field measurements; and 4) dynamics of sediment transport. This volume provide valuable information to engineers, hydrologists and other technical personnel dealing with flood risk analysis and protection against floods.

SCIENTIFIC COMMITTEE

FOREWORD

This specialty conference has been organized by ENEL spa and CNR-GNDCI, under the sponsorship of ASCE; its aim is to establish the state of the art in mathematical modelling of flood propagation over initially dry areas. In this respect the conference could provide valuable information to engineers, hydrologists and other technical personnel dealing with flood risk analysis and protection against floods.

The need for reliable evaluation tools of flood risk potential is nowadays paramount both in developed and developing countries, although the motivations are different in the two types of situation. As in developed countries flood plains have already been occupied by urban settlements, industrial activities and extensive infrastructures, the applicable studies are mainly addressed to the verification of flood defence works (embankments, dams, etc.). On the other hand, in developing countries mathematical modelling can be used for development planning and improvement, in order to avoid future high risk situations.

The problem of flood analysis has recently found a renewed interest in Italy, due to recent regulations requiring the determination of areas located downstream of dams subject to flooding in the events of opening of the outlet works and collapse of the dam.

Although initially focused to the specific topic of 2-D modelling of flood propagation over initially dry areas, the subjects of the conference have been successively broadened to other closely related topics, such as algorithmic and computational aspects of modelling and the use of Digital Terrain Models and Geographical Information Systems in flood modelling.

The participating papers have been divided into the four following homogeneous groups:

MODELLING OF FLOOD PROPAGATION

1. Theoretical and numerical aspects of flood modelling
2. Computational and practical aspects. Use of G.I.S. and other computer based tools
3. Model verification against laboratory and field measurements
4. Dynamics of sediment transport

Researchers and technical personnel interested in 2-D shallow water modelling can find a variety of up-to-date contributions on the many different aspects of this scientific field.

Paolo Molinaro
Luigi Natale

Milano, May 2nd, 1994

C O N T E N T S

Session 2: Computational and Practical Aspects.
Use of G.I.S. and other computer based tools

Session 3: Model verification against laboratory
and field measurements

Session 4: Dynamics of Sediment Transport

Theoretical and Numerical

Aspects of flood modelling

COMPUTING TWO DIMENSIONAL FLOOD PROPAGATION WITH A HIGH RESOLUTION EXTENSION OF MCCORMACK'S METHOD

F. Alcrudo[1] and P. García Navarro

Departamento de Ciencia y Tecnología de Materiales y Fluídos
Facultad de Ciencias, Universidad de Zaragoza. 50009 - Zaragoza, SPAIN
[1] On leave at Department of Mechanical & Aeronautical Engineering, University of California Davis. Davis, CA-95616, USA.

ABSTRACT

McCormack's method, as applied for solving the bidimensional de Saint Venant equations is enhanced with the aid of the theory of Total Variation Diminishing (TVD) schemes in order to obtain an accurate, robust time marching algorithm in finite volume formulation for calculating the progressing of flood waves in two dimensional areas.

INTRODUCTION

From the fluid mechanics point of view flood propagation is an extremely complicated phenomenon involving the dynamics of a fluid with a free boundary in intense turbulent motion under the acceleration of gravity. When trying to describe mathematically this situation, one is faced with solving a full three dimensional unsteady Navier-Stokes problem with a free boundary. It is well known, however, that this is not actually feasible if all the space and time scales involved during a flood are to be resolved, at least with current algorithms and computers.

If only the main features of the flow pattern are of interest and no attention is paid in resolving smaller scale effects such as secondary flows, boundary layers or turbulence, one may resort to a simpler mathematical representation of the physical reality. This is certainly the case in many engineering applications, where depending on the cause that produced the flood, its severity or the time interval to be modelled, a simple kinematic description can be sufficient to provide the requested answers.

In this paper we shall be concerned with situations in which the full nonlinear de Saint Venant or Shallow Water equations are needed to account for dynamical effects such as the propagation of water fronts or bores. One of the reasons that renders the solution of the Shallow Water equations difficult is precisely the appearance of such bores since they mathematically correspond to discontinuous or weak solutions of the differential equations.

Very often the numerical solution of de St. Venant's system is tackled by means of shock capturing algorithms, also known as through methods, which automatically produce and track the evolution of any discontinuity in the solution. Despite the conceptual simplicity of through methods, they are subject to many deficiencies that need to be handled with care if an accurate representation of the flow is sought. Appearance of unwanted numerically generated oscillations whenever a shock is formed, and the possibility of computing non-physical discontinuities are two of the main drawbacks usually tailored to classical shock capturing schemes. It is worth noting here that these effects have not only an impact on the accuracy and reliability of the solution, but in many cases can seriously impair the computational procedure due to their destibilizing effect.

Classical methods can usually been enhanced with various kinds of artificial viscosity terms in order to prevent those problems to happen. However artificial viscosity can cure the problem only to a limited extent and moreover it must be tuned for every considered problem. When computational models have to be calibrated, as it is frequently the case in free surface flows, having to deal with extra numerical parameters can be cumbersome and misleading. Sometimes one can read how terms relating to turbulent eddy viscosity and numerical viscosity play an indistinguishable role in practical computations.

McCormack's scheme is among the most popular algorithms for solving the type of problem considered here. Being second order accurate in space and time it provides enough resolution of the flow field and has been successfully used to solve the Shallow Water equations in two dimensions (Fennema and Chaudry [1], Bento Franco and Betamio [2]). However it is not free of the problems quoted earlier as can be seen in the above referenced papers.

Newly developed algorithms have successfully overcome the undesired features of classical schemes and have been applied to bore propagation problems. On the light of these more elaborated techniques some classical schemes can be reformulated in a way that allows for considerable improvement of their performance at a reduced computational expense and reprograming effort. Total Variation Diminishing (TVD) methods, though not certainly perfect have proven to be much superior in many types of flow problems than classical ones. As regards to free surface flows, evidence of these superiority can be found in the works by Alcrudo [3], Alcrudo and Garcia Navarro [4], [5].

Based on the theory of TVD methods, self adaptive numerical dissipation terms are added to McCormack's scheme that prove very effective in suppressing spurious oscillations and in discarding non-physical solutions of the Shallow Water system. The terms added make use of flux difference splitting and nonlinear limiter functions to automatically provide the appropriate amount of numerical viscosity around sharp variations of the variables without affecting regions of smooth flow. As regards practical implementation of the method, the flexible finite volume technique is preferred in order to deal easily with non-cartesian meshes without the use of coordinate transformations.

DE SAINT VENANT MODELS IN TWO DIMENSIONS

It is assumed that the flow is mainly two dimensional taking place in a horizontal plane (x-y) parallel to the earth surface, and being described by the water depth, h, and the two cartesian components of the water velocity in the plane of motion, u, v. Any dependence of the flow variables on the vertical coordinate (z) is neglected and it is also assumed that the vertical velocity is zero. Considering a hydrostatic pressure distribution, the 2-D de Saint Venant equations can be derived by depth averaging the Navier-Stokes equations with appropriate boundary conditions. In this work the terms arising from viscous and turbulent internal stresses are not taken into account for simplicity and compactness. Their numerical treatment does not pose any special problem (save for the implementation of an adequate turbulence model) and can be accomplished with central second order differences. Surface friction stresses with the bottom have been included by means of empirical formulae such as Manning's.

Written in general conservative form the de Saint Venant equations in two dimensions read:

$$\frac{\partial \mathbf{U}}{\partial t} + \nabla \mathbf{F} = \mathbf{H} \tag{1}$$

$\mathbf{U}$ is the vector of flow variables, $\mathbf{F}$ the momentum flux tensor and $\mathbf{H}$ represents the source term of momentum due to bed slope and bottom friction. More commonly, expressing $\mathbf{F}$ in its cartesian components, $\mathbf{E}$, $\mathbf{G}$, Equation 1 is written:

$$\frac{\partial \mathbf{U}}{\partial t} + \frac{\partial \mathbf{E}}{\partial x} + \frac{\partial \mathbf{G}}{\partial y} = \mathbf{H} \tag{2}$$

The vectors $\mathbf{U}$, $\mathbf{E}$ and $\mathbf{G}$ can be expressed in terms of the flow variables as:

$$\mathbf{U} = \begin{pmatrix} h \\ hu \\ hv \end{pmatrix} \qquad \mathbf{E} = \begin{pmatrix} hu \\ hu^2 + gh^2/2 \\ huv \end{pmatrix} \qquad \mathbf{G} = \begin{pmatrix} hv \\ huv \\ hv^2 + gh^2/2 \end{pmatrix} \tag{3}$$

where g stands for the gravity acceleration.

The source term, **H** is:

$$\mathbf{H} = \begin{pmatrix} 0 \\ gh\,(S_{ox} - S_{fx}) \\ gh\,(S_{oy} - S_{fy}) \end{pmatrix} \qquad (4)$$

S_{ox} and S_{oy} represent the bed slopes in the two cartesian directions, and S_{fx} and S_{fy} the friction slopes, that can be calculated from Manning's formula:

$$S_{fx} = \frac{u\,n^2\,\sqrt{u^2 + v^2}}{h^{4/3}} \;, \quad S_{fy} = \frac{v\,n^2\,\sqrt{u^2 + v^2}}{h^{4/3}} \qquad (6)$$

where n is Manning's roughness coefficient.

FINITE VOLUME DISCRETIZACION

To allow the simple use of irregular (non cartesian) grids adapted to complex geometrical boundaries, the spatial discretization is done in a cell-centered finite volume formulation. The two dimensional physical domain is decomposed into elementary quadrilaterals and their centroids labelled with a pair of subindexes following the usual ordering for the mesh coordinates. Every quadrilateral will be considered an elementary control volume, its surface boundary being formed by the four straight sides or walls enclosing it. Obviously in two dimensions volumes are not true volumes, but rather surface areas, and boundaries to these volumes are not true surfaces but rather closed lines. Nevertheless the terms volumes and surfaces enclosing them are used here in an attempt to keep the nomenclature general.

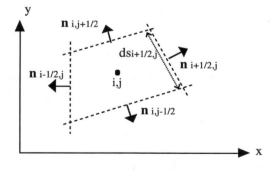

Figure 1: Description of an elementary control volume

Figure 1 shows one of those elementary control volumes, labelled (i,j). It is surrounded by four walls w_r (r=1,4). Each one of these walls is located at a half index distance from the centroid of the volume and thus the following anti-clockwise convention can be taken: $w_1=(i+1/2,j)$, $w_2=(i,j+1/2)$, $w_3=(i-1/2,j)$, $w_4=(1,j-1/2)$. The generalized surface of any wall is in fact its length, represented in the figure by the symbol ds_{w_r}. The unit vector normal to each wall is also displayed under the symbol n_{w_r}.

Equation 1 can be integrated over an arbitrary control volume V as:

$$\int_V \frac{\partial U}{\partial t}\, dV + \int_V \nabla F\, dV = \int_V H\, dV \qquad (7)$$

or applying Gauss theorem to the second integral:

$$\int_V \frac{\partial U}{\partial t}\, dV + \oint_S \left(F \cdot n\right) ds = \int_V H\, dV \qquad (8)$$

being S the surface enclosing V.

Denoting by U_{ij} and H_{ij} the average value of the flow variables and the source term respectively over the control volume V_{ij}, at a given time, Equation 8 can be rewritten:

$$\frac{\partial U_{i,j}}{\partial t} = \frac{-1}{V_{i,j}} \oint_{S_{i,j}} \left(F \cdot n\right) ds + H_{i,j} = RHS_{i,j} \qquad (9)$$

The quantity RHS_{ij} being the residual in cell (i,j), defined here only for the sake of compactness.

In order to integrate Equation 9 in time to obtain the evolution of the flow, the surface integral must be evaluated first. In this work it is approximated as the sum over the four walls surrounding volume (i,j) of a numerical flux function, F^*, which is dependent upon the scheme chosen:

$$\oint_{S_{i,j}} \left(F \cdot n\right) ds \approx \sum_{r=1}^{4} \left(F_{w_r}^* \cdot n_{w_r}\right) ds_{w_r} \qquad (10)$$

Equation 10 is applied to every cell of the computational domain, thus obtaining the residual for every control volume. Once all the residuals are known, one can integrate Equation 9 by an appropriate method. In case of McCormack's scheme, Equation 10 must be evaluated twice every time step for each cell because it uses a predictor-corrector time integration. Denoting the time step as Δt and the time level with a superindex, McCormack's scheme reads in condensed notation as:

$$U_{i,j}^p = U_{i,j}^n + \Delta t \cdot RHS_{i,j}^n$$

$$U_{i,j}^c = U_{i,j}^n + \Delta t \cdot RHS_{i,j}^p \qquad (11)$$

$$U_{i,j}^{n+1} = \frac{1}{2}\left(U_{i,j}^p + U_{i,j}^c\right)$$

where p and c stand for predictor and corrector values. The numerical flux function to be used in Equation 10 is different at the predictor and corrector stages. Taking for instance the wall $w_1=(i+1/2,j)$ one has for the predictor evaluation of the residual:

$$\left(\mathbf{F}_{w_1}^* \cdot \mathbf{n}_{w_1}\right)^n = \left(\mathbf{F}^* \cdot \mathbf{n}\right)_{i+1/2,j}^n = \mathbf{F}_{i+1,j}^n \cdot \mathbf{n}_{i+1/2,j} \tag{12}$$

while for the corrector evaluation:

$$\left(\mathbf{F}_{w_1}^* \cdot \mathbf{n}_{w_1}\right)^p = \left(\mathbf{F}^* \cdot \mathbf{n}\right)_{i+1/2,j}^p = \mathbf{F}_{i,j}^p \cdot \mathbf{n}_{i+1/2,j} \tag{13}$$

The expressions for the other three walls enclosing cell (i,j) are found by simply shifting the indexes in Equations 12 and 13 to the ones of the corresponding wall.

Other equally valid version of the method would have been found by just switching the predictor and corrector definitions. In fact four different variants can be written for McCormack's scheme in two dimensions and it is well known that all the four versions have to be alternated in a symmetric sequence in order to avoid directional biases of the results.

HIGH RESOLUTION EXTENSION

Methods capable of rendering monotone front profiles and discarding nonphysical solutions while being second order accurate are known as high resolution methods. In this section TVD diffusive terms are added to the algorithm described above in order to obtain a high resolution extension of McCormak's scheme. Basically the procedure is a finite volume formulation extension to two dimensions of the one reported in García Navarro and Alcrudo [6]. A detailed derivation of the method can be found in Alcrudo [3].

An artificial diffusion flux tensor $\mathbf{D}$, to be defined later, is introduced such that Equation 8 is modified to:

$$\int_V \frac{\partial \mathbf{U}}{\partial t} \, dV + \oint_S \left(\mathbf{F} \cdot \mathbf{n}\right) ds - \oint_S \left(\mathbf{D} \cdot \mathbf{n}\right) ds = \int_V \mathbf{H} \, dV \tag{14}$$

The discretization of the second surface integral in the right hand side of Equation 14 is performed as in Equation 10:

$$\oint_{S_{i,j}} \left(\mathbf{D} \cdot \mathbf{n}\right) ds \approx \sum_{r=1}^{4} \left(\mathbf{D} \cdot \mathbf{n}\right)_{w_r} ds_{w_r} \tag{15}$$

and the scheme is modified by replacing only the updating step in Equation 11 with:

$$\mathbf{U}_{i,j}^{n+1} = \frac{1}{2}\left(\mathbf{U}_{i,j}^p + \mathbf{U}_{i,j}^c\right) + \Delta t \left[\sum_{r=1}^{4} \left(\mathbf{D} \cdot \mathbf{n}\right)_{w_r} ds_{w_r}\right]^n \tag{16}$$

The definition of the projection of the diffusion flux tensor along the normal to the selected wall, $(\mathbf{D}\cdot\mathbf{n})_{w_r}$, is based on the jacobian matrix, $\mathbf{J}_{w_r}$, of the projection of the flux tensor $(\mathbf{F}\cdot\mathbf{n})_{w_r}$ along the same normal:

$$\mathbf{J}_{w_r} = \frac{\partial(\mathbf{F}\cdot\mathbf{n})_{w_r}}{\partial \mathbf{U}} \tag{17}$$

To fix ideas, suppose that we are dealing with wall $w_1=(i+1/2,j)$. An approximate jacobian matrix $\widetilde{\mathbf{J}}_{i+1/2,j}$ is sought, such that it has real and distinct eigenvalues, and a complete set of eigenvectors. Moreover it is required that if we denote by $\Delta(x)_{i+1/2,j}$ the jump in any quantity x across the $(i+1/2,j)$ interface, i.e. $\Delta(x)_{i+1/2,j}=(x)_{i+1,j} - (x)_{i,j}$, the following property be fulfilled (Roe [7]):

$$\Delta(\mathbf{F}\cdot\mathbf{n})_{i+1/2,\ j} = \widetilde{\mathbf{J}}_{i+1/2,j} \cdot \Delta(\mathbf{U})_{i+1/2,\ j} \tag{18}$$

Such approximate jacobian matrix can be calculated by assuming that its structure is similar to that of the true jacobian $\mathbf{J}_{i+1/2,j}$ and then imposing Equation 18. The result is:

$$\widetilde{\mathbf{J}}_{i+1/2,j} = \begin{pmatrix} 0 & n_x & n_y \\ (\widetilde{c}^2-\widetilde{u}^2)n_x-\widetilde{u}\widetilde{v}n_y & 2\widetilde{u}n_x-\widetilde{v}n_y & \widetilde{u}n_y \\ -\widetilde{u}\widetilde{v}n_x+(\widetilde{c}^2-\widetilde{v}^2)n_y & \widetilde{v}n_x & \widetilde{u}n_x+2\widetilde{v}n_y \end{pmatrix} \tag{19}$$

being $n_{i+1/2,j}=(n_x,n_y)$ and the overbar quantities defined as follows:

$$\widetilde{u} = \frac{u_R \sqrt{h_R} + u_L \sqrt{h_L}}{\sqrt{h_R} + \sqrt{h_L}}$$

$$\widetilde{v} = \frac{v_R \sqrt{h_R} + v_L \sqrt{h_L}}{\sqrt{h_R} + \sqrt{h_L}} \tag{20}$$

$$\widetilde{c} = \sqrt{\frac{g(h_R + h_L)}{2}}$$

where subindex R stands for $(i+1,j)$ and L for (i,j). It is interesting to write the eigenvalues of $\widetilde{\mathbf{J}}_{i+1/2,j}$:

$$\widetilde{a}^1 = \widetilde{u}n_x + \widetilde{v}n_y + \widetilde{c}$$
$$\widetilde{a}^2 = \widetilde{u}n_x + \widetilde{v}n_y \tag{21}$$
$$\widetilde{a}^3 = \widetilde{u}n_x + \widetilde{v}n_y - \widetilde{c}$$

and its eigenvectors:

$$\widetilde{\mathbf{e}}^1 = \begin{pmatrix} 1 \\ \widetilde{u} + \widetilde{c}n_x \\ \widetilde{v} + \widetilde{c}n_y \end{pmatrix} \quad \widetilde{\mathbf{e}}^2 = \begin{pmatrix} 0 \\ -\widetilde{c}n_y \\ \widetilde{c}n_x \end{pmatrix} \quad \widetilde{\mathbf{e}}^3 = \begin{pmatrix} 1 \\ \widetilde{u} - \widetilde{c}n_x \\ \widetilde{v} - \widetilde{c}n_y \end{pmatrix} \tag{22}$$

Now one can expand the jump $\Delta(\mathbf{U})_{i+1/2,j}$ into the eigenvalues of $\widetilde{\mathbf{J}}_{i+1/2,j}$ as:

$$\Delta(\mathbf{U})_{i+1/2,j} = \sum_{k=1}^{3} \widetilde{\alpha}^k \widetilde{\mathbf{e}}^k \tag{23}$$

the coefficients are:

$$\tilde{\alpha}^{1,3} = \frac{\Delta h}{2} \pm \frac{1}{2\tilde{c}}\left[\Delta(hu)\,n_x + \Delta(hv)\,n_y - \left(\tilde{u}n_x + \tilde{v}n_y\right)\Delta h\right]$$

$$\tilde{\alpha}^2 = \frac{1}{\tilde{c}}\left[\left(\Delta(hv) - \tilde{v}\Delta h\right)n_x - \left(\Delta(hu) - \tilde{u}\Delta h\right)n_y\right]$$

(24)

where the subindex (i+1/2,j) has been dropped from any $\Delta()$ for clarity.

With all the definitions above, the normal diffusion flux, $(\mathbf{D}\cdot\mathbf{n})_{i+1/2,j}$, can be expressed as:

$$\left(\mathbf{D}\cdot\mathbf{n}\right)^n_{i+1/2,\,j} = \frac{1}{2}\sum_{k=1}^{3}\alpha^k\,\psi\!\left(\tilde{a}^k\right)\left[1 - \lambda\left|\tilde{a}^k\right|\right]\left[1 - \varphi(\gamma^k)\right]\tilde{e}^k$$

(25)

where λ is a measure of the mesh ratio: $\lambda = \Delta t/d_{i+1/2,j}$, $d_{i+1/2,j}$ being the distance between centroids (i,j) and (i+1,j).

$\psi(a)$ is an entropy correction to the modulus of a, that avoids the appearance of non-physical solutions. In its simplest formulation it just replaces lal by a small number (of order 1) whenever lal is smaller than the number itself. Other formulae can be found in Hirsch [8].

Finally φ is a limiter function responsible for obtaining monotone solutions. It depends on, γ^k, the ratio of a product of certain quantities in neighboring walls:

$$\gamma^k_{i+1/2,\,j} = \frac{\left[\alpha\,\psi(\tilde{a})\left(1 - \lambda\,|\tilde{a}|\right)\right]^k_{i+1/2-s,\,j}}{\left[\alpha\,\psi(\tilde{a})\left(1 - \lambda\,|\tilde{a}|\right)\right]^k_{i+1/2,\,j}} \quad ; \quad s = \text{sign}\!\left(\tilde{a}^k_{i+1/2,j}\right)$$

(26)

Several expressions for φ are available in the cited references. In the examples computed in this paper, minmod function was used.

$$\text{minmod}(\gamma) = \max[\,0, \min(1,\gamma)\,]$$

(27)

Numerical stability is assured by the following restriction in the time step:

$$\Delta t \leq \min_{i,j}\left\{\frac{\left(d_{i\pm 1/2,j\pm 1/2}\right)}{2\left(\sqrt{u^2 + v^2} + c\right)_{i,j}}\right\}$$

(28)

The method described above needs a word of explanation. It must be first stated that true TVD methods only exist for scalar equations in one space dimension (1-D). Nevertheless there is overwhelming numerical evidence that their extension to systems of equations in 1-D produce high resolution results. McCormack's predictor-corrector method is made TVD-like by adding specific artificial diffusion terms that were in fact devised for Lax-Wendroff (one step) methods (Alcrudo [3]). For linear equations both schemes are identical and indeed so they are their TVD versions. But when one deals with nonlinear equations, Lax-Wendroff and McCormack's algorithms are no longer the same and thus one can not assure the TVD properties for the latter. The analysis can not be carried through because of the nonlinearity coupled with the predictor-corrector sequence. Despite these facts, numerical experiments show that very good results are usually produced.

As regards the two-dimensional case, the only theoretical result available (Leveque 1990 [9]) says that no 2-D scheme of second order accuracy can be TVD. Again experiments show, however, that results produced with 2-D extensions of TVD-type 1-D schemes are of much better quality than those of classical methods.

Also caution must be exercised when the mesh used contains severe distortions and skewness. Only for regular or smoothly varying grids will the computed results be reliable.

COMPUTED EXAMPLES

The results from two known flood propagation testcases are discussed in this section. In both cases initial conditions of zero velocity everywhere and specified water depth in the domain considered are imposed. Then the flow is allowed to evolve according to de Saint Venant equations subject to appropriate boundary conditions. A detailed description of the treatment of boundary conditions is not intended in this paper. Nevertheless it can be said that they are imposed using the theory of characteristics. Only the information corresponding to bicharacteristics entering the domain of integration is imposed while that linked to outgoing bicharacteristics is extrapolated from interior points to the boundaries. In case of solid walls the tangency condition is enforced along the cells touching the material border.

Fennema and Chaudry testcase

Fennema and Chaudry [1] computed the formation of a two dimensional front in a 200m x 200m flat bed, frictionless, rectangular reservoir. Initially the reservoir was divided in two halves separated by a dam. The half to the left of the wall contained water 10m deep while the one to the right contained water only 5m deep. At the initial instant a portion of the dam 85m wide was removed, allowing water to flow from the left to the right. Those initial conditions are not very demanding to the numerical method because they lead to subcritical flow everywhere in the domain. In this work we preferred to try higher values of the depth ratio thus decreasing the initial depth in the downstream side to 1m. The mesh is the same as the one used by the original authors (cartesian $\Delta x=\Delta y=5m$). A perspective view of the free surface at t=5s after the removal of the dam can be seen in Figure 2a as calculated with classical McCormack method without artificial viscosity. It can be said that the results are completely spoiled by

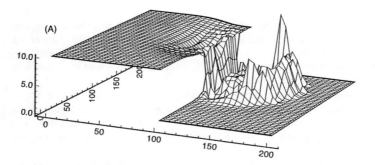

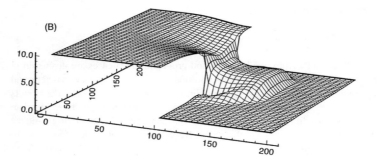

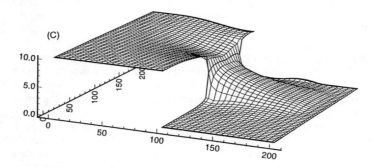

Figure 2: Perspective of the free surface for Fennema and Chaudry testcase. a) Initial depth ratio of 10, original McCormack scheme b) Initial depth ratio of 10, McCormack TVD method. c) Initial depth ratio of 1000, McCormack TVD method.

instabilities in the form of wide amplitude wiggles. The solution computed by means of McCormack TVD version is shown on Figure 2b, where the improvement is easily noticed.

Figure 2c shows the results computed with McCormack TVD method for an initial depth of 0.01m at the downstream side at t=5s after the breach was opened. No interpolation of the water surface at the initial instant was performed (such as the very used Ritter profile), but just an abrupt discontinuity in the water depths each side of the breach was allowed to evolve numerically. It is worth noting the stability and robustness of the proposed method since it is well known that calculations started with effective discontinuous initial conditions usually blow up during the early stages.

Bellos, Soulis and Sakkas [10] experiment

These authors provided experimental data of the evolution of a dam break induced wave propagating in a converging diverging flume (21m x 1.4m) as a model of a valley. Data for wet and dry initial beds, as well as for different slopes are given in their work. Manning's roughness is kept fixed at n=0.012. Basically water stages at six measuring stations along the flume are provided as a function of time.

In this work only two cases are shown: the first one corresponds to zero bed slope, 0.3m initial depth upstream of the dam and 0.053m downstream. It is to be remarked again that initial conditions for the numerical calculation were truly discontinuous. The grid used is the one provided by the authors in their original work (44 x 8 cells). A comparison of the calculated depth (solid line) with experimental values (circles) versus time is shown in Figure 3. Only the four sections which compare less favorably are shown for space economy. Section number 2 is upstream of the dam, section 3 is at the dam site and the other two are downstream of it. The wave reflected from the downstream end is not well reproduced by the calculation because open boundary conditions are difficult to simulate numerically. However, intensity and time of arrival of the dam break wave are predicted correctly. In the two sections not shown very good agreement between measurements and calculation is found.

The second case regards initial dry bed downstream of the dam and bed slope $S_{ox}=0.01$, $S_{oy}=0$. Truly dry bed is difficult to handle by any numerical method. Manning's friction term has to be switched off in regions of zero or very small depth to prevent division by zero. On the other hand any small amplitude oscillation can drive the depth to negative values thus giving rise to inconsistencies. For the dry bed computation reported here the depth is reset to a very small value (of the order of 10^{-4}) any time it falls below the threshold. Also flow velocity is then set to zero. The comparison between experimental and computed results can be seen in Figure 4, for the four sections that compare less favorably. Overall agreement can be considered satisfactory.

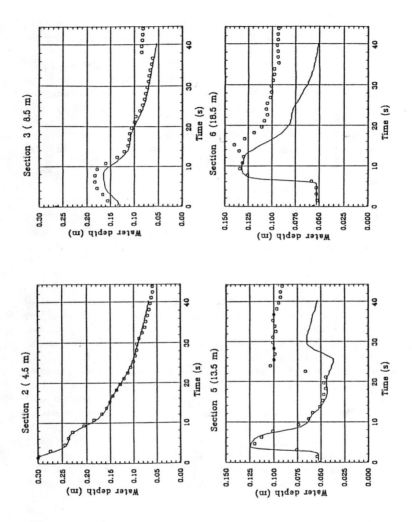

Figure 3: Bellos et al. experiment. Computed (—) and measured (o) water depth versus time at four flume locations. $S_{ox}=S_{oy}=0$. Wet bed initial conditions.

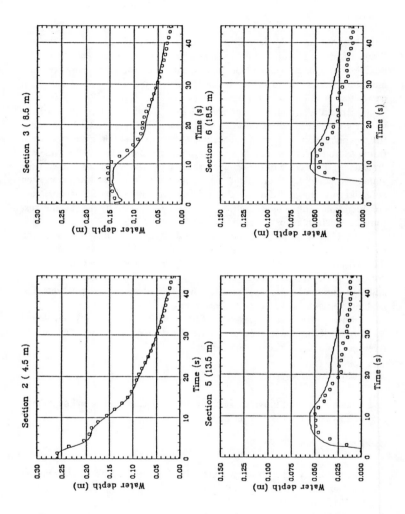

Figure 4: Bellos et al. experiment. Computed (—) and measured (o) water depth versus time at four flume locations. $S_{ox}=0.01$, $S_{oy}=0$. Dry bed initial conditions.

CONCLUSIONS

The well known McComack's method has been applied to the full Shallow Water equations in finite volume formulation. This technique leads to a more flexible spatial discretization than the original finite differences. However care must be taken so as to preserve smoothness in the variation of the mesh parameters in order to avoid accuracy losses.

Artificial dissipation terms derived from Total Variation Diminishing schemes have been added in order to produce sharp representation of water fronts while retaining the original accuracy of the method in regions of smooth flow. Although no second order accurate algorithm can be TVD in two dimensions, the ability of the resulting scheme to cope with complicated flow situations including supercritical conditions and regime transitions has been considerably enhanced over the original method. Besides theoretical considerations, strong evidence for these ideas is provided by the calculations, some of which are reported in this paper.

It is to be remarked that in contrast with classical artificial viscosities, the numerical dissipation added here contains no adjustable terms or parameters, thus avoiding the time consuming trial and error process needed to tune a program for every particular problem. The level of numerical viscosity is adjusted automatically, high at discontinuities to prevent oscillations and negligible elsewhere to produce accurate results.

Implementing the proposed modifications into any already developed McCormack-based code should produce satisfactory gains in accuracy and robustness at a reduced computational cost.

ACKNOWLEDGEMENTS

Financial support provided for this work by Diputación General de Aragón is gratefully acknowledged by the authors.

REFERENCES

1. Fennema, R.J. and Chaudry, M.H. Explicit Methods for 2-D Transient Free surface Flows. Journal Of the Hydraulics Division, ASCE, Vol.116, pp.1013-1034, 1990.
2. Bento Franco, A. and Betamio, A. Simulaçao Uni e Bidimensional de Cheias Provocadas por Roturas de Barragens em Planicies de Inundaçao. V Simposio Luso-Brasiliero de Hidraulica e Recursos Hidricos, SILUSB, 1991.

3. Alcrudo, F.   Esquemas de Alta Resolución de Variación Total Decreciente para la Simulación de Flujos Discontínuos de Superficie Libre. Ph.D. Thesis, Universidad de Zaragoza, 1992.
4. Alcrudo, F., García Navarro, P. and Savirón, J.M.   Flux Difference Splitting for the 1D Open Channel Flow Equations. International Journal for Numerical Methods in Fluids, Vol. 14, pp. 1009-1018, 1992.
5. Alcrudo, F. and García Navarro, P.   A High Resolution Godunov-Type Scheme in Finite Volumes for the 2D Shallow-Water Equations. International Journal for Numerical Methods in Fluids, Vol. 16, pp.489-505, 1993.
6. García Navarro, P. and Alcrudo, F.   1-D Open Channel Flow Simulation using TVD McCormack scheme, Journal of Hydraulic Engineering, ASCE, Vol.118, No.10, pp.1359-1372, 1992.
7. Roe, P.L.   Approximate Riemann Solvers, Parameter Vectors and Difference Schemes, Journal of Computational Physics, No. 43, pp. 357-372, 1981.
8. Hirsch, Ch.   Numerical Computation of Internal and external Flows. Vol.2: Computational Methods for Inviscid and Viscous Flows, Wiley, Chichester, 1990.
9   Leveque, R.J.   Numerical Methods for conservation Laws, Birkhäuser-Verlag, Basel, 1990.
10. Bellos, C.V., Soulis, J.V. and Sakkas, J.G.   Computation of two-dimensional dam-break induced flows, Advances in Water Resources. Vol.14, No.1, pp.31-41, 1991.

A 2D numerical simulation of the Po river delta flow

D. Ambrosi, CRS4, Via Nazario Sauro, 10, 09123, Cagliari, Italy
S. Corti, V. Pennati, ENEL-DSR-CRIS, Via Ornato 90/14, 20162, Milano, Italy
F. Saleri, Politecnico di Milano, P.le Leonardo da Vinci 32, 20133, Milano, Italy

Abstract

A numerical simulation of the steady flow in the proximity of the delta of the Po river has been carried out. We consider the 2D quasi-hydrostatic model, approximating non conservation form of the shallow water equations by two new different finite element schemes. In the present paper we describe the differences among the two discretizations, focusing the advantages of each of them. Then we present another finite element model solving the conservation form of shallow water equations. Then we introduce the physical problem and the characterics of the flow we want to simulate. Finally we compare and discuss the numerical results, at the light of the theory underlying the schemes. The present study is intended as a preliminary step towards the study of the real evolution of the unsteady flow as governed by the inflow and the outflow tidal elevation evolving in time.

1   Introduction

Although a fully physically–consistent description of a river flow can be achieved only by a three dimensional approach, the implementation of an efficient and accurate 2D model is still an interesting task. In fact, most of the mathematical and numerical difficulties of a 3D model are already present in its 2D counterpart and, moreover, the 2D case has to be exploited to minimize the CPU requirements that become huge in the 3D case. A 3D code that is expected to solve a real flow problem in a parallel architecture, should rely on a 2D model which runs on a workstation in a few hours.

[0]Work supported by ENEL-CRIS and the Sardinian Regional Authorities

In the present paper we consider two new finite element schemes to approximate the shallow water equations (SWE); the main difference among the codes is the way in which they advance in time. The first approach we consider (SWEET1) solves the system equations one by one, and the coupling between the components is ensured by an iterative procedure. In this way, the CPU requirements are substantially reduced, because a system one-third large is to be solved, but the implicit coupling between the equations has to be checked *a posteriori*. The second approach (SWEET2) has a more physical basis; it nearly resembles the method by Benquè et al. [3] , although traduced in the FE framework. The main idea is to split the equations at a differential level at every time step, in order to decouple the physical wave contributions. In particular, the wave travelling at speed $\sqrt{gh}$, which is known to be the most restrictive for what regards the time step in this kind of problems, is treated implicitly together with the continuity equation. The latter method is the less expensive than the former, requiring only the inversion of a symmetric matrix at each time step. However, the split equations that are solved at every time step may have a different nature of the original ones, and consistency is not *a priori* ensured. The paper is organized as follows:

- in section 2 the SWE are introduced,

- the new numerical methods are described in section 3,

- the MONOS method is described in section 4,

- in section 5 the geometry and boundary conditions of the physical problem are described,

- in section 6 the numerical results are presented and discussed.

2 The shallow water equations

The SWE in primitive differential form read

$$\frac{\partial \mathbf{v}}{\partial t} + (\mathbf{v} \cdot \nabla)\mathbf{v} - \nabla \cdot (\mu \nabla \mathbf{v}) + g\nabla \xi = -g\frac{\mathbf{v}|\mathbf{v}|}{hC^2}, \tag{1}$$

$$\frac{\partial \xi}{\partial t} + \nabla \cdot (h\mathbf{v}) = 0, \tag{2}$$

where $\mathbf{v}(x, y, t) = (v_x, v_y)^T$ represents the velocity vector, ξ is the elevation over a reference plane, h is the total depth, μ is the turbulent horizontal diffusion coefficient, g is the gravity acceleration and C is the Chezy coefficient. A schematic representation of the involved quantities may be seen in fig.1. In the present study we omit to consider Coriolis force and wind stress, as these terms have a minor importance for the class of problems we are dealing with.

According to the characteristic theory, as long as the flow is subcritical two boundary conditions are to be prescribed at the inflow and one at the outflow, plus a further condition everywhere, acconting for the second order term in the momentum equation. The former condition used in the present study will be described later, the latter is satisfyied applying a weak Neumann condition on v, which arises naturally in the integration by parts of the diffusive term. The wall boundary conditions we consider here are free slip boundary conditions, i.e. the velocity component normal to the boundary is imposed to be null, whilst the tangential component is computed as in rest of the domain.

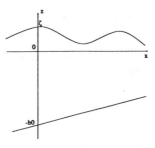

Figure 1: Elevation and depth.

3 The new numerical methods

The space discretization of equations (1-2) is based on a Galerkin finite element method. The details of the Galerkin approach we use may be found in [1] and [2]. The main properties are the following:

- the interpolation functions are taken in P1 for elevation, P1 or P2 for the velocity; as usual, P1 is the set of the piecewise linear functions, P2 the set of the piecewise quadratic functions.

- the stabilization of the convective term in the momentum equation is achieved in two ways: by adding a dissipative term, analogously to the SUPG procedure for the advection–diffusion equation [6], or by operating a lagrangian fractional step [7]. Both stabilizations are consistent with the primitive form of the equations we have adopted. The advantages of each formulation are well known: a lagrangian fractional step allows a larger time step, but it is computationally more expensive and produces more numerical dissipation; moreover, in order to get a satisfactory representation of the pathline, at least quadratic elements should be used.

- we use the ILU preconditioned CG or GMRES method to invert the stiffness matrices resulting from the discretization [5].

- free slip wall boundary conditions are imposed at an algebraic level requiring that the normal projection of the left hand side operator appearing in equation (1) is zero.

Two ways to advance in time are currently used and will be described in the following subsections.

3.1   The SWEET1 approach

The first approach we consider is a classical first order fully implicit scheme discretizing the non conservation form of the SWE:

$$\frac{\mathbf{v}^{n+1} - \mathbf{v}^n}{\Delta t} + (\mathbf{v}^{n+1} \cdot \nabla)\mathbf{v}^{n+1} - \nabla \cdot \mu\nabla\mathbf{v}^{n+1} + g\nabla\xi^{n+1} = -g\frac{\mathbf{v}|\mathbf{v}|}{hC^2}, \quad (3)$$

$$\frac{\xi^{n+1} - \xi^n}{\Delta t} + h^{n+1}\nabla \cdot \mathbf{v}^{n+1} + \mathbf{v}^{n+1} \cdot \nabla h^{n+1} = 0. \quad (4)$$

To solve the nonlinear problem that arises at each time step, we have proposed in [2] the following iterative procedure:

$$\frac{\mathbf{v}_m^{n+1} - \mathbf{v}^n}{\Delta t} + (\mathbf{v}_{m-1}^{n+1} \cdot \nabla)\mathbf{v}_m^{n+1} - \nabla \cdot \mu\nabla\mathbf{v}_m^{n+1} \quad (5)$$

$$-\frac{\mathbf{v}_m^{n+1}|\mathbf{v}_{m-1}^{n+1}|}{h_{m-1}^{n+1}C^2} = -g\nabla\xi_{m-1}^{n+1},$$

$$\frac{\xi_m^{n+1} - \xi^n}{\Delta t} + h_m^{n+1}\nabla \cdot \mathbf{v}_m^{n+1} + \mathbf{v}_m^{n+1} \cdot \nabla h_m^{n+1} = 0, \quad (6)$$

where $m = 1, 2, \ldots$ is an inner iteration counter.
At each iteration we solve two scalar advection-diffusion equations for the velocity components and a transport equation for the elevation. Note that the momentum equation here is decoupled into two independent scalar equations. We have analyzed the stability properties of this scheme in [2]. The iterative procedure stops when convergence is achieved.

3.2 The SWEET2 approach

The main idea behind our second approach is to split the equations at a differential level at every time step, in order to decouple the physical wave contributions. At every fractional step the equations to be solved are:

step 1

$$\mathbf{v}^{n+1/3} - \mathbf{v}^n = \Delta t \left[-(\mathbf{v}^n \cdot \nabla)\mathbf{v}^n \right], \qquad (7)$$

step 2

$$(h\mathbf{v})^{n+2/3} - (h\mathbf{v})^{n+1/3} = \Delta t \left[\nabla \cdot \left(\mu h^n \nabla \mathbf{v}^{n+1/3} \right) - g\frac{\mathbf{v}^{n+1/3}|\mathbf{v}^{n+1/3}|}{C^2} \right], \quad (8)$$

step 3

$$(h\mathbf{v})^{n+1} - (h\mathbf{v})^{n+2/3} + \Delta t\, gh^{n+1}\nabla\xi^{n+1} = 0, \qquad (9)$$
$$\xi^{n+1} - \xi^n + \Delta t\, \nabla \cdot (h\mathbf{v})^{n+1} = 0. \qquad (10)$$

At the first step the advective part of the momentum equation is integrated in primitive form; to ensure consistency with the conservative form used in the other steps, an extra term should be added in (9) (see [3]), that we neglect for its minor practical importance. The non-conservative form of eq.(7) allows to use stable and efficient discretizations as the lagrangian method or SUPG. At the third step the two equations are decoupled taking the divergence of the first one and subtracting. It is finally found to be solved the following Helmholtz–type equation:

$$\xi^{n+1} - \xi^n - (\Delta t)^2 \nabla \cdot \left(gh^{n+1}\nabla\xi^{n+1} \right) = -\Delta t \, \nabla \cdot (h\mathbf{v})^{n+2/3}. \qquad (11)$$

The new elevation is therefore used to solve equation (9).

The main advantage of this fractional step procedure is that the wave travelling at speed $\sqrt{gh}$, which is the most restrictive for what regards the time step in this kind of problems, is in some sense decoupled and treated implicitly. Thefore the CFL condition given by celerity $\sqrt{gh}$ is economically circumvented, as only a symmetric matrix has to be inverted at every time step.

The whole scheme may be made more implicit taking the diffusive term in the momentum equation at time $n + 1$; this may be useful when large turbulent diffusion coefficients are used.

It should be remarked that the two approaches do not differ only in the time advancing method (in this case they should coincide for a steady problem), because the fractional step procedure here adopted implies a decoupling of the equations term at each time step that is not present in the original SWE system. Some considerations about this issue may be found in [1].

4 The MONOS approach

The MONOS model is based on the two-dimensional SWE solved in the conservation form:

$$\frac{\partial P}{\partial t} + \frac{\partial}{\partial x}\left(\frac{P^2}{h} + \frac{gh^2}{2}\right) + \frac{\partial}{\partial y}\left(\frac{PQ}{h}\right) - \frac{\partial}{\partial x}T_{xx} - \frac{\partial}{\partial y}T_{xy} = \qquad (12)$$
$$-\frac{gP}{C^2 h^2}(P^2 + Q^2)^{\frac{1}{2}}$$

$$\frac{\partial Q}{\partial t} + \frac{\partial}{\partial x}\left(\frac{PQ}{h}\right) + \frac{\partial}{\partial y}\left(\frac{Q^2}{h} + \frac{gh^2}{2}\right) - \frac{\partial}{\partial x}T_{xy} - \frac{\partial}{\partial y}T_{yy} = \qquad (13)$$
$$-\frac{gQ}{C^2 h^2}(P^2 + Q^2)^{\frac{1}{2}}$$

$$\frac{\partial h}{\partial t} + \frac{\partial P}{\partial x} + \frac{\partial Q}{\partial y} = 0 \qquad (14)$$

where P and Q are the unit-width discharges in the x and y directions respectively, h is the water depth and T_{xx}, T_{xy} and T_{yy} are the components of the tensor of the internal unit-width forces due to turbolence:

$$T_{xx} = 2h\mu\frac{\partial}{\partial x}\left(\frac{P}{h}\right); \quad T_{xy} = h\mu\left[\frac{\partial}{\partial x}\left(\frac{Q}{h}\right) + \frac{\partial}{\partial y}\left(\frac{P}{h}\right)\right]; \quad T_{yy} = 2h\mu\frac{\partial}{\partial y}\left(\frac{Q}{h}\right) \quad (15)$$

The space discretization of equations (12)-(16) is based on a Galerkin finite element method in which polynomial shape functions are of first order for the water depth and second order for the discharges. The time integration is performed according to the θ-method where, to respect stability properties, θ must lie within [0.5,1.]. In the numerical experiments related to the Po river we adopted $\theta=0.7$. To solve the non-linear problem that arises at each time step, in MONOS the equations are solved iteratively by means of the Newton-Raphson technique,and at each iteration the resulting system of linear equations is solved by means of a Gaussian elimination method, optimized according to the frontal technique. Free slip wall boundary conditions are imposed requiring that boundary forces are zero where discharges are unknown. A detailed description of the model can be found in [4].

5 The Po river delta

In the present study we consider the delta of the Po river, i.e. the last 20 km. of its lenghth. A plot of the river contours together with a very rough description of its depth variations may be seen in fig.2.

The inflow is 379 m. large, the depth varies between 17 m. and few centimeters. Below the first bifurcation the river has a sharp elbow; here

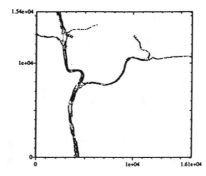

Figure 2: Contour of the Po river delta

the river width diminishes and depth grows strongly. Here the velocity **v** and especially the unit-width discharge $h\mathbf{v}$ are espected to assume large values.

The boundary conditions we consider are aimed to our final task, which is to simulate a flow evolution as governed by the tidal elevation variation. Therefore we impose the elevation everywhere in the open boundary; moreover, if a boundary segment is found to be of inflow type, the tangential velocity component is prescribed to be null. Considering the river open boundaries starting from the bottom in fig.2 and considering the open boundaries in a conterclockwise sense, the imposed elevations are 0.46 m., 0.30 m., 0.30 m., 0.41 m., 0.32 m., 0.38 m.

We have used two different meshes to compute the flow. The coarse (P2) one has 7528 nodes and 3185 elements. The fine (P1) one has 10602 elements and 6481 nodes, with a minimum triangle diameter of 8 meters. The code SWEET1 has been applied to the former using a P2P1 representation for velocity and elevation respectively, SWEET2 has been applied to the latter using a P1 representation for both the unknowns. A detail of the fine mesh in two crucial zones may be seen in fig.3; the choosen regions correspond to the main bifurcation of the river and the the first bifurcation of the right branch.

6   Numerical results

For the present computation the Chezy coefficient has been set equal to 50 $m^{1/2}s^{-1}$ for the computations with all codes. The viscosity coefficient, that should account for turbulence and vertical dishomogeneities, has been set equal to 50 m^2s^{-1}.

In fig.(4)-(6) are shown the velocity vectors as computed by the codes

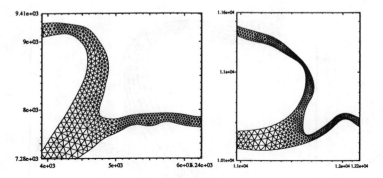

Figure 3: A detail of the fine mesh in two crucial zones

in the same regions where the mesh has been detailed. In all plots there not appear oscillations in the velocity field, in accordance with the smooth results we have for elevation. Oscillations are absent also where the velocity has large gradients, as in the upper part of fig.5; the elevation has there a variation of 5 centimeters.

Conclusions

The code SWEET has been described in its main features and has been applied to the simulation of the Po river steady flow at its delta. The accuracy and stability properties that had been shown in test cases of previous papers seem to be confirmed in this real flow problem. The results obtained by MONOS seem to be physically acceptable and are comparable with those obtained by SWEET. The comparison with experimental data in the case of varying tidal elevation at the boundary will give a quantitative measure of its accuracy.

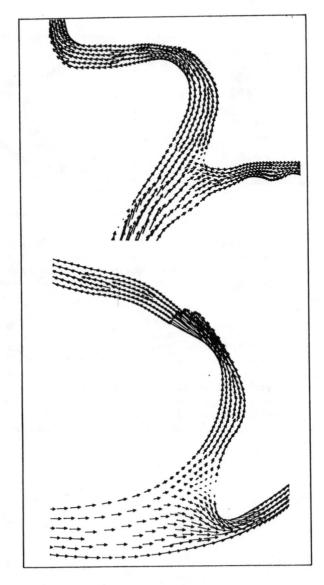

Figure 4 : Zooms of the velocity field by SWEET1

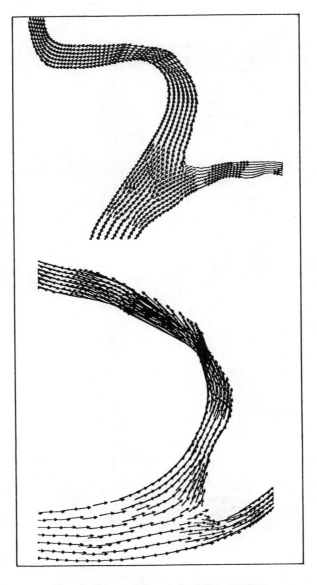

Figure 5 : Zooms of the velocity field by SWEET2

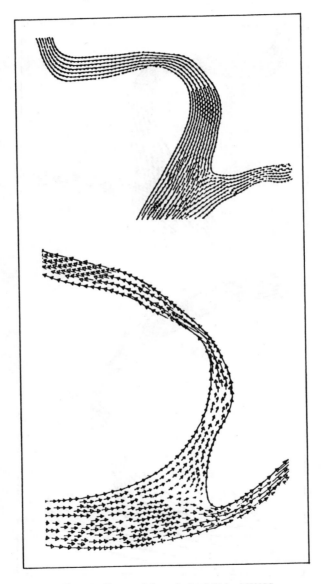

Figure 6 : Zooms of the velocity field by MONOS

BIBLIOGRAPHY

[1] V.I. Agoshkov, D. Ambrosi, V. Pennati. A. Quarteroni, F. Saleri, *Mathematical and numerical modelling of shallow water flow*, Computational Mechanics, 11, 280–299 (1993)

[2] V.I. Agoshkov, E. Ovchinnikov, A. Quarteroni, F. Saleri, *Recent Developments in the Numerical Simulation of Shallow Water Equations. II Temporal Discretization*, to appear in Mathematical Models and Methods in Applied Sciences.

[3] J.P. Benquè, J.A. Cunge, J. Feuillet, A. Hauguel, and F.M. Holly, *New method for tidal current computation*, Journal of Waterway, Port, Coastal and Ocean Division ASCE, 108, 396–417 (1982)

[4] A.Di Monaco, P. Molinaro, *A Finite element 2-dimensional model of free surface flows: verification against experimental data of the problem of the emptying of a reservoir due to dam-breaking*, Int.Conf. on Computer Methods and Water Resources, Rabat, 14 March (1988).

[5] G.H. Golub, C.F. Van Loan, *Matrix computations*, The John Hopkins University Press, Baltimora (1989)

[6] T.J.R. Hughes, *Finite element methods for fluids*, AGARD Report 787 on "Unstructured Grid Method for Advection Dominated Flows" (1992)

[7] O. Pironneau, *On the Transport–Diffusion Algorithm and its Application to the Navier–Stokes Equations*, Numerische Matematik, 38, 309–332 (1982)

An Adaptative Nine-point Finite Volume Roe Scheme for Two-Dimensional Saint-Venant Equations

Fayssal Benkhaldoun

Luc Monthe

LMI, INSA BP 8, 76131 Mont-Saint-Aignan cedex, France

ABSTRACT

The system of Saint Venant equations is considered numerically. A Roe approximate solver coupled to a nine-point finite volume spatial discretisation is introduced at the same time as a mesh adaptation procedure. The successful simulation of a dam breaking proves the robustness of the scheme.

1. INTRODUCTION

The system of Saint Venant equations has been studied in many ways in the last decade. Simulation of water phenomena is, indeed, of great importance for the prediction of dam breaking or flood propagation effects as well as sea and estuary pollution state.

Since the system of equations exhibits a hyperbolic part physically related to convection and acoustic wave propagation, one has to carefully choose the numerical scheme to insure numerical stability as well as good accuracy.

The existing codes are of two kinds, those based on the method of the characteristics (see HERVOUET and all [1]), and those based on predictor-corrector type schemes like the Mac Cormack one (see KETTANI [2]).

Unfortunately in both cases, conservation of the schemes cannot be proven.

The purpose of the scheme we are presenting here is to handle this aspect of the numerical simulation difficulties.

Seeking to carry out simulations in significantly deformed areas, with the use of mesh adaption, we exclude orthogonal meshes. The finite volume discretisation is then a set of arbitrary quadrilaterals, this makes the use of a classical five-point spatial discretisation not suitable, mainly if one wants to integrate diffusion terms (see FAILLE [3]). Hence, a nine-point finite volume scheme was chosen. Furthermore, an explicit time integration is used in order to keep down the memory use. Since the centered spatial discretisation of the convective terms is numerically instable when used together with explicit time integration, we have chosen a non centered scheme based on the Roe approximate solution of local Riemann problems, which ensures numerical stability under certain CFL conditions. Finally, to increase the accuracy of the computations, a strategy of dynamic contraction of the mesh in the high gradient regions was utilized.

2. THE EQUATIONS TO BE SOLVED

The set of conservative equations to be solved is the following:

$$\frac{\partial h}{\partial t} + \frac{\partial hu}{\partial x} + \frac{\partial hv}{\partial y} = 0$$

$$\frac{\partial hu}{\partial t} + \frac{\partial (hu^2 + \frac{g}{2}h^2)}{\partial x} + \frac{\partial huv}{\partial y} = gh(S_{ox} - S_{fx})$$

$$\frac{\partial hv}{\partial t} + \frac{\partial huv}{\partial x} + \frac{\partial (hv^2 + \frac{g}{2}h^2)}{\partial y} = gh(S_{oy} - S_{fy})$$

where h is the water depth, u an v the velocity components. g is the gravity acceleration, S_{ox} and S_{oy} the slopes of the canal in the x and y direction, and S_{fx} and S_{fy} are friction terms

This system of equations can be written:

$$\frac{\partial W}{\partial t} + \frac{\partial (F(W))}{\partial x} + \frac{\partial (G(W))}{\partial y} = S(W) \tag{1}$$

3. MATHEMATICAL APPROACH

One can formally write the problem we are studying, under the form: Find $W \in C^0([0,T], H[\Omega])$ such that:

$$\frac{\partial W}{\partial t} + \frac{\partial (F(W))}{\partial x} + \frac{\partial (G(W))}{\partial y} - S(W) = 0 \quad \text{in } \Omega,$$

$$\alpha W + \beta \frac{\partial W}{\partial n} = g \quad \text{over } \Gamma,$$

$$W(X,0) = W^0(X) \quad \text{in } \Omega$$

where Ω is a bounded open set of IR^2, $\Gamma = \partial\Omega$, and $H(\Omega)$ is a hilbert space, here $L^2(\Omega)$.

For such a complicated set of equations, it is rather difficult to derive a solution in the strong meaning or even to prove the existence and uniqueness of this solution. Nevertheless, we know that physical solutions do exist, so we seek here to compute a weak solution of the problem by first writing an approximation of the temporal derivation which gives: For $W^n \in L^2(\Omega)$, find $W^{n+1} \in L^2(\Omega)$ such that:

$$W^{n+1} = W^n + \Delta t A(W^n) \tag{2}$$

where A is an operator including the transport terms of the PDE and the source terms. Since we are not sure that those terms are differentiable in the classical sense, we consider the equation (2) in the sense of distribution which is equivalent to writing:

$$< W^{n+1}, \phi >_{L^2(\Omega)} \;=\; < W^n + \Delta t A(W^n), \phi >_{L^2(\Omega)} \quad \forall \phi \in D'(\Omega) \tag{3}$$

To get an approximation of W^{n+1}, we project (3) in a finite dimension subspace V_h of $L^2(\Omega)$. So we rewrite our problem:

For $W^n \in V_h(\Omega)$ find $W^{n+1} \in V_h(\Omega)$, such that:

$$< W^{n+1}, \phi_i >_{L^2(\Omega)} = < W^n + \Delta t A(W^n), \phi_i >_{L^2(\Omega)} \tag{4}$$

For every ϕ_i where $(\phi_i)_{i=1,...,M}$ is a basis of $V_h(\Omega)$

More precisely, we consider a partition of Ω in quadrilateral cells C_i, $i = 1,..,nc$ and take the basis of $V_h(\Omega)$, (χ_i) $i = 1,..,nc$, where χ_i is the caracteristic function of the cell C_i.

The equation (4) can be now written as:

$$< W^{n+1}, \chi_i >_{L^2(\Omega)} \quad = \quad < W^n + \Delta t A(W^n), \chi_i >_{L^2(\Omega)} \qquad i = 1,..,nc$$

4- SPATIAL DISCRETISATION AND STABILITY REQUIREMENTS

Equation (1) is of the evolution kind, it exhibits a convective part (F and G) and a source part (S).

Let us consider a simpler analogous linear scalar equation, with a convective term and a source term.

$$U_t + aU_x = cU \tag{5}$$

We aim to analyse the numerical stability if one uses a forward Euler discretisation for the temporal term, and a centered spatial discretisation for the convective term. We use the classical Fourier analysis and inject the mode $\hat{U}_j^n = \lambda^n e^{ij\theta}$ in equation (5), we get:

$$\lambda^2 = [1 + c\Delta t]^2 + (\frac{a\Delta t}{\Delta x})^2 sin^2\theta$$

one easily note that if $c = 0$ (such is the case when the convective aspect is predominant in the phenomenon, for example a dam breaking in a horizontal canal without friction), $\lambda^2 > 1$ when $\theta \neq k\pi$, and the scheme is inconditionaly unstable.

On the contrary, if one utilizes a decentered scheme for the convective term (backward if $a > 0$ for example), one obtains:

$$\lambda^2 = [1 + c\Delta t - 2\frac{a\Delta t}{\Delta x}sin^2\frac{\theta}{2}]^2 + (\frac{a\Delta t}{\Delta x})^2 sin^2\theta$$

and it is easier to derive stability conditions in this case.

Spatial discretisation of the transport term

We shall utilize what one calls an upwind finite volume discretisation of the convective part of equation (1). We consider the reduced equation :

$$W_t + F(W)_x + G(W)_y = 0 \qquad (6)$$

A great amount of litterature has been written about the numerical handling of this equations (see VAN LEER[4], GOUDONOV [5], LAX [6]), and we shall present only the mean stages of the scheme we have written. The main idea then, is to integrate equation (6) over a finite volume, here the cell C_i, which gives, with the choice of V_h, and making use of Green's formula:

$$\text{Aire}(C_i)\frac{\partial W_i}{\partial t} + \int_{\partial C_i} F(W)n_x + G(W)n_y d\sigma = 0$$

one has then to evaluate the flux $\quad F(W)n_x + G(W)n_y$ over the four borders of the cell $\quad C_i$.

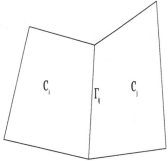

Construction of the ROE scheme: Let us write:

$$IF(W,\vec{n}) = n_x F(W) + n_y G(W)$$

we seek an approximation:

$$\int_{\Gamma_{ij}} IF(W,\vec{n})d\sigma \;=\; \Phi(W_{ij}^l, W_{ij}^r, \vec{n}_{ij})\, meas(\Gamma_{ij})$$

where Φ is called the numerical flux, and the states W_{ij}^l and W_{ij}^r are interpolated by the nine points finite volume scheme as follows:

- determine the middle point M of Γ_{ij}

- determine the points I^l and I^r

- interpolate W_{ij}^l between $W(G_i)$ and $W(G_l)$

- interpolate W_{ij}^r between $W(G_j)$ and $W(G_k)$

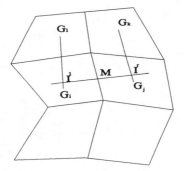

Roe (see ROE [7]) suggested a particular choice of Φ based upon the resolution of approximate linear Riemann problems:

$$\Phi(W_{ij}^l, W_{ij}^r, \vec{n}_{ij}) \;=\; \frac{1}{2}[IF(W_{ij}^l, \vec{n}_{ij}) + IF(W_{ij}^r, \vec{n}_{ij})] \tag{1}$$

$$-\; \frac{1}{2}|A^*(W_{ij}^l, W_{ij}^r, \vec{n}_{ij})|(W_{ij}^r - W_{ij}^l) \tag{2}$$

with the following requirements about the matrix A^*:

$i) A^*(U,V,\vec{n})(V-U) \;=\; IF(V,\vec{n}) - IF(U,\vec{n})$

$ii) A^*(U,U,\vec{n}) = A(U,\vec{n}) = \frac{\partial}{\partial u}IF(U,\vec{n})$

$iii)A^*(U, V, \vec{n})$ has real eigenvalues for all the states U and V and the eigenvectors of A^* are linearly independant

Property i) ensures the conservation of the scheme, while the consistency with the original problem is garanteed by property ii), and the hyperbolicity by property iii).

construction of $A^*(W^l, W^r, \vec{n})$:   Following former results (see ALCRUDO [8]), we introduce the parameter vector:

$$W^*(W^l, W^r) = \{h^*, h^*u^*, h^*v^*\}^T$$

where

$$h^* = \frac{1}{2}(h^l + h^r),$$

$$u^* = \frac{\sqrt{h^l}u^l + \sqrt{h^r}u^r}{\sqrt{h^l} + \sqrt{h^r}},$$

$$v^* = \frac{\sqrt{h^l}v^l + \sqrt{h^r}v^r}{\sqrt{h^l} + \sqrt{h^r}},$$

and define

$$A^*(W^l, W^r) = A(W^*(W^l, W^r)).$$

One can notice that:

$h^* \to h$ when $W^l \to W^r \to W$

$u^* \to u$ when $W^l \to W^r \to W$

$v^* \to v$ when $W^l \to W^r \to W$

which implies:

$W^*(W^l, W^r) \to W$ when $W^l \to W^r \to W$

and then,

$A^*(W^l, W^r) = A(W^*(W^l, W^r)) \rightarrow A(W)$ when $W^l \rightarrow W^r \rightarrow W$

and an algebraic computation enables us to verify that

$$A^*(W^l, W^r)(W^r - W^l) = IF(W^r) - IF(W^l)$$

the two conditions i) and ii) are then fulfilled and the condition iii) is obvious since $A^* = A(W^*)$.

Spatial discretisation of source terms: We have to evaluate the term

$$\int_{C_i} S(W)dv$$

This term is simply discretised by writing

$$\int_{C_i} S(W)dv = \text{ Area }(C_i)S(W_i)$$

5. MESH ADAPTATION STRATEGY

Since we are studying a transient phenomenon with moving sharp gradient regions, we aim to contract the mesh around those regions without adding more nodes. The idea is to consider the mesh like a connected network of springs, such that the stiffness K_{ij} of a spring bordered by the two nodes M_i and M_j is related to the value of the numerical solution on the two nodes.

One can then write the strength between M_i and M_j:

$$\vec{F}_{ij} = K_{ij} \, \vec{M_iM_j}.$$

By writing that each node is in equilibrium one gets two linear systems:

$$AX = Bx$$
$$AY = By$$

where X and Y are the vectors of the coordinates of the mesh, A is a square matrix related to the springs' stifness, and Bx and By are related to the conditions imposed on the mesh within the limits of the computational domain. In the present work, we chose:

$$K_{ij} = 0.5 \left(grad\, h_i + grad\, h_j \right)$$

where $grad\, h_i$ is the spatial gradient of the water depth in the node M_i.

Mesh control and interpolation

Seeking to avoid overlapping of the mesh, we restrict the motion of each node M_i in a circle centered in M_i and which ray is the distance from M_i to his nearest neighbour. If one note M_i^{new} the position of the node M_i after adaption and control, the solution in M_i^{new} is linearly interpolated by a barycentric formula involving M_i^{new} and the four connected neighbours of M_i in the former mesh.

6. NUMERICAL EXPERIMENTS

The first experiment we performed is the simulation of a dam breaking in a rectangular horizontal canal without friction terms. To evaluate the efficency of the scheme, we compared the numerical results to the self similar analytical solution which is composed of a rarefaction wave followed by a shock. The initial water depth is 5 m to the left of the dam and 1 m to the right, with a zero initial velocity everywhere in the canal. The results are plotted at time t=0.64 s.

A first run was performed with a mesh adaptation occuring at time t=0.46s (Fig 1, 3 and 5b). In a second run two adaptations were done, the first at time t1 = 0.39 s and the second at time t2 = 0.47s (Fig 2, Fig 4, Fig 5c and Fig 5d). One can notice the good behavior of the Roe scheme, and the great improvement of the results under the effect of mesh adaptation.

7. CONCLUSION

In the present work, we developped a robust solver for the system of Saint Venant equations based on Roe approximate Riemann solver. The stiff situation of a dam breaking was simulated with great success, and an appreciable improvement is introduced by the use of mesh adaptations. Following this first attempt, the work should progress towards a more accurate solver based on second order spatial discretisation, and on an improvement of the Roe scheme for the handling of situations where one has dry areas and stiff friction terms.

REFERENCES

1. Hervouet, J. M. and Watrin, A. Code TELEMAC (Systeme Ulysse) Résolution des équations de Saint Venant bidimensionnelles, EDF report HE-43/87.37.

2. Ech-Cherif El Kettani B., Berrada A., Ouazar D., Agousoul M., 2D Dam-Break Flood-Wave Propagation on dry beds, 2nd Int. Conf. CMWR, Rabat, 7-11 Oct, 1991.

3. Faille, I., Les volumes finis en vue de la modélisation bidimensionnelle de la génèse et de la migration des hydrocarbures, Institut Français du pétrole P, RB20, RR 37763, Géologie 29655, Projet 4137, Etude 4137003.

4. Van Leer, B. Flux vector splitting for the Euler equations, Lecture Notes in Physics, 170 (Springer, Berlin, 1982).

5. Godunov, S. K. A finite difference method for the numerical computation of the discontinuous solutions of the equations of flow dynamics. Mat. Sb., v 47, 1959, pp 271-290.

6. Lax, P. D. Hyperbolic systems of conservation laws and the mathematical theory of shock waves. SIAM Regional Conference series in App. Math., 11. 1973.

7. Roe, P. L. Approximate Riemann Solvers, Parameter vectors, and difference scheme. Journal of Computational Physics 43, 357-372, 1981.

8. Alcrudo, F., Garcia-Navaro, P., Saviron, J. M., Flux difference splitting for 1D open channel flow equations. Int. J. Num. Meth. Fluids, Vol 14, 1009-1018 (1992).

Fig. 1. Water depth at t=0.64s (no Adaption(.) Adaption(o) Exact(-))

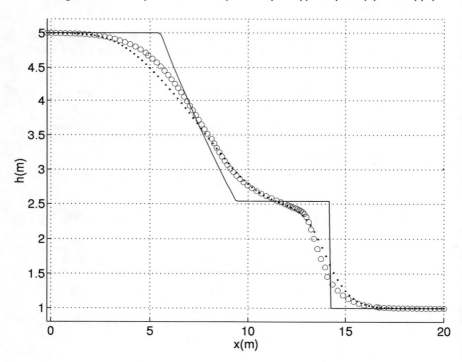

Fig. 2. Water depth at t=0.64s (no Adaption(.) Adaption(o) Exact(-))

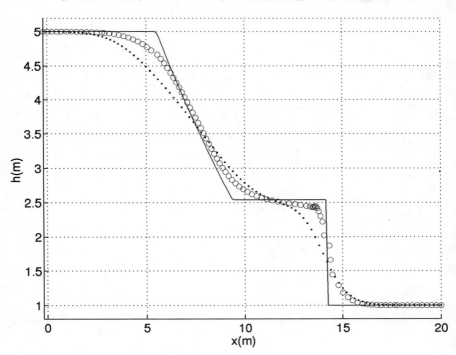

Fig. 3. Flow velocity at t=0.64s (no Adaption(.) Adaption(o) Exact(-))

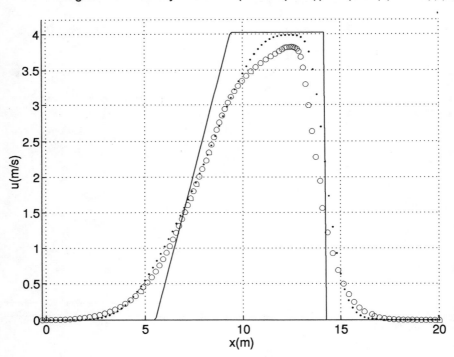

Fig. 4. Flow velocity at t=0.64s (no Adaption(.) Adaption(o) Exact(-))

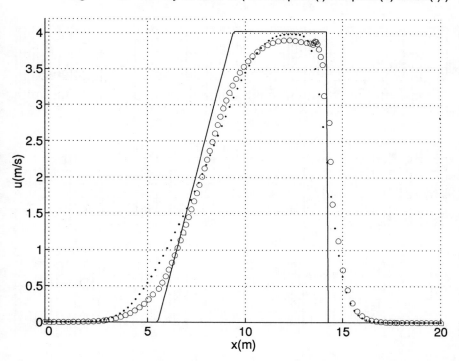

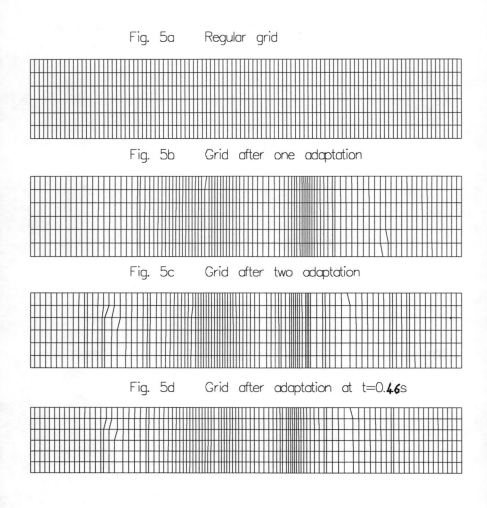

Fig. 5a   Regular grid

Fig. 5b   Grid after one adaptation

Fig. 5c   Grid after two adaptation

Fig. 5d   Grid after adaptation at t=0.46s

Plain flooding: near field and far field simulations.

G. Braschi, F. Dadone, M. Gallati
Hydraulic and Environmental Engineering Department
University of Pavia - Italy

ABSTRACT

The paper deals with the problems met in the attempt of simulating the plain submersion produced by a levee breaking in the inertia dominated region (near field), scaling with the typical breach dimension, and in the far field, scaling with the typical plain length, where the flow is essentially controlled by the bottom topography.
A fractional step algorithm based on a 1D decomposition strategy and just designed to deal with transcritical flows is presented.
It is employed to get the near field simulation of the plain flooding due to a sudden wall gate opening; the results are compared to laboratory data.
Some far field real world simulations are finally presented.

INTRODUCTION

The aim of the present work is to report about the studies and the experience in plain submersion simulation performed at the Department of Hydraulic and Environmental Engineering of Pavia University. The interest for such a subject sprang from a research project sponsored by the Italian Research Council group on hydrogeological disaster prevention (GNDCI), aiming at the definition of suitable procedures to draw rational and consistent flooding scenarios produced by dam or levee breaking or bank overtopping of a river, either in country or urban environment.
The main part of this paper is devoted to the description of the algorithm developed by the authors to obtain the numerical solution of the wholly conservative 2D flow equations. Its performance is pointed out by comparison with laboratory data obtained for the purpose.

In the second part the approximations adopted for the simulation of large uneven areas submersion are shortly reviewed and some results are reported.

LOCAL AND GLOBAL SCALES IN A LEVEE BREACH PLAIN FLOODING

The flow characteristics of the plane flooding due to a point-like source, say a levee breach, are quite different in regions close to the breach and far away. In the 'near' field, say in a region scaling with the typical breach dimension lenght, the flow is essentially controlled by the inertial effects: it can become locally supercritical and then revert to the typical subcritical condition with an approximately circular (often undular) jump.
On the contrary, in the 'far' field, characterized by the typical lenght scale of the plane, the ground topography, roughness and soil imbibition, are likely to be the most important factors in determining the flow characteristics. It is very difficult, if not impossible, to realistically describe the advancement of the wet front at the local scale in presence of natural topography. It is well recognized, in any case, that mainly the soil properties control the flow when the water depth is small enough, so that, in far-field computations, it is reasonable to neglect convection inertial effects everywhere including the front region.

NEAR FIELD SIMULATION

It is admitted that in a region close to the breach, scaling with its typical horizontal lenght, the inertial effects are responsible of the main flow characteristics.
The mind picture leading to the 'near field' definition is at least rather crude when compared to the real world phenomenon. It is quite reasonable to think that the mechanism of the breaking strongly affects the development of the flow in its very first stages and presumably the local unpredictable geometry modification essentially controls the motion which is likely to be 3D.
The problem is studied with reference to simplified 'breach hypoteses' with the aim of getting possible rational pictures of the flooding close enough to the breach: they are expected to be useful in defining the source for the subsequent large field simulation, where the direction of the flow is controlled by the presence of local obstacles.
It is well known that strong convection effects cause either supercritical or subcritical flow with the occurrence of shocks in the transition from the first to the second condition. The numerical

algorithm should automatically deal with such situations detecting the shocks and providing the meaningful physical force and mass balance across them.

In the last 15 years several numerical schemes meeting the above requirements have been proposed and applied with considerable success to mathematically similar problems of gasdynamics [1] and several attempts have been made to employ the same techniques for the hydraulic simulation of rapidly varied shallow water flow problems [2]. The subject is now well explored in the 1D case but still needs some investigation effort and experience in the 2D situations.

The hydraulic problems differ from the gasdynamic ones mainly for the presence of the bottom source terms (accounting for slope and friction effects) and because of the existence of a lower limit for the water depth (technically relevant in the submersion problems).

Flow mathematical description

The rapidly varied unsteady flow in shallow waters is described, under well known hypotheses, e.g. [3] and [4], by a differential system that can be written in strictly conservative form as :

$$\frac{\partial U}{\partial t} + \frac{\partial F(U)}{\partial x} + \frac{\partial G(U)}{\partial y} = S \qquad (1)$$

$U = [\, h, q_x, q_y \,]$; $F(U) = [\, q_x, \Sigma_x, M \,]$; $G(U) = [\, q_y, M, \Sigma_y \,]$

$S = [\, 0 \,, gh(S_{0x}\text{-}S_{fx}) \,, gh(S_{0y}\text{-}S_{fy}) \,]$

$\Sigma_x = uq_x + gh^2/2$; $\Sigma_y = vq_y + gh^2/2$; $M = vq_x = uq_y$

The symbols have the following meaning: t [s] time; x,y [m] cartesian space coordinates; g [m/s^2] gravity acceleration ; h [m] water depth over the field bottom; q_x, q_y [m^2/s] unit-width discharge components; u,v [m/s] mean velocity components; Σ_x, Σ_y[m^3/s^2] unit-width total momentum components; M[m^3/s^2] unit-width cross momentum; S_{0x} , S_{0y} bottom slope in x and y direction; S_{fx}, S_{fy} friction slope in x and y direction to be computed with the Manning extended formulas:

$$S_{fx} = n^2 \frac{(u^2+v^2)^{1/2}}{h^{1.33}} u \quad ; \quad S_{fy} = n^2 \frac{(u^2+v^2)^{1/2}}{h^{1.33}} v$$

The differential problem is correctly posed once the above equations are supplied with the proper initial and boundary conditions. The hyperbolic nature of the system is highlighted by the existence of two systems of characteristic surfaces [5]: following them the proper number of boundary conditions can be set at the flow entrance and exit boundaries, both in the subcritical and supercritical case, as well as at the solid boundaries.

Numerical conservative integration scheme.
The steps that bring to the numerical solution algorithm of the above system are shortly described in the following.
The procedure is seen as the logical extension of the technique already employed and largely tested in rapidly varied monodimensional flow problems described in [6] : in this case the characteristic form of the equations is quite simple and a strategy of solution oriented with the local characteristic directions and based on the compatibility equations is readly got.

In view of the numerical solution, let us first split in time the system (1) into the two following 1D-like equivalent ones after introducing the intermediate time evaluation h' :

$$\frac{h'-h_0}{\delta t} + \frac{\partial q_x}{\partial x} = 0 \quad ; \quad \frac{q_x - q_{x0}}{\delta t} + \frac{\partial \Sigma_x}{\partial x} = gh(S_{0x}-S_{fx}) - \frac{\partial M}{\partial y} \quad (2a)$$

$$\frac{h-h'}{\delta t} + \frac{\partial q_y}{\partial y} = 0 \quad ; \quad \frac{q_y - q_{y0}}{\delta t} + \frac{\partial \Sigma_y}{\partial y} = gh(S_{0y}-S_{fy}) - \frac{\partial M}{\partial x} \quad (2b)$$

The solution h^n, q_x^n, q_y^n at time t+dt, starting from time t, can be got solving two 1D problems linked by the momentum source-like cross derivatives.
Let us choose an explicit strategy to discretize the space derivatives: the choice is justified by the small time scales of the transients to be simulated.

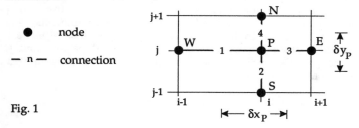

●   node

— n —   connection

Fig. 1

We state to evaluate the space derivatives (and the sources) with the most updated available values of the variables that we suppose stored in the nodes of a cartesian network (Fig. 1).

A conservative solution algorithm needs the evaluation of the 'star fluxes' (q_x^*, q_y^*, Σ_x^*, Σ_y^*, M^*) in the reaches connecting two adjacent nodes.

With reference to the symbols of Fig. 1, the local algebraic approximate forms of the differential equations (2a),(2b) are:

$$\frac{h'-h}{\delta t} + \frac{q_{x3}^*-q_{x1}^*}{\delta x_P} = 0 \; ; \; \frac{q_x^n-q_x}{\delta t} + \frac{\Sigma_{x3}^*-\Sigma_{x1}^*}{\delta x_P} = gh_P\,(S_{0xP}-S_{fxP}) - \frac{M_4^*-M_2^*}{\delta y_P} \quad (3a)$$

$$\frac{h^n-h'}{\delta t} + \frac{q_{y4}^*-q_{y2}^*}{\delta y_P} = 0 \; ; \; \frac{q_y^n-q_y}{\delta t} + \frac{\Sigma_{y4}^*-\Sigma_{y2}^*}{\delta y_P} = gh_P\,(S_{0yP}-S_{fyP}) - \frac{M_3^*-M_1^*}{\delta x_P} \quad (3b)$$

The star quantities are obtained from the available values according to an upwind strategy as follows.

Consider, for example, the horizontal connection "1" between points W and P where the values $q_{x1}^*, \Sigma_{x1}^*, M_1^*$ are to be computed.

First we estimate the reach variables u_1, v_1 and $c_1 = \sqrt{gh_1}$ through simple averages of the flow variables h, q_x, q_y to get:

$$h_1 = \frac{h_W + h_P}{2} \; ; \; u_1 = \frac{u_W h_W + u_P h_P}{h_W + h_P} \; ; \; v_1 = \frac{v_W h_W + v_P h_P}{h_W + h_P}$$

so that the reach Froude number, $Fr_1 = u_1/c_1$, is defined.

If the flow is locally considered as unidimensional and the source terms are neglected, the hyperbolic system (2a) can be written in the well known characteristic form:

$$dx/dt = u \pm c \qquad\qquad (4a)$$
$$dq_x = -(c \pm u)\,dh \qquad\qquad (4b)$$

meaning that the compatibility conditions (4b) hold along the characteristic lines with slope (4a).

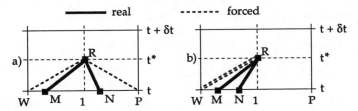

Fig. 2 - Characteristics: a) subcritical case, b) supercritical case

According to the value of the reach Froude number Fr_1, when u_1 is positive, the two situations depicted in Fig. 2 a) and b) occur in the subcritical case and in the supercritical one respectively.

Let us first consider the subcritical case. We 'approximate' the characteristic directions RM, RN with the lines RW and RP. Then the extrapolated reach values h_R, q_{xR} taken say in point R, are readly computed by means of the compatibility equations (4b) discretized as:

$$q_{xR}-q_{xW} = -(c_1+u_1)\,(h_R-h_W) \quad ; \quad q_{xR}-q_{xP} = -(c_1-u_1)\,(h_R-h_P)$$

This step brings to the required value of $q_{x1}^* = q_{xR}$:

$$q_{x1}^* = \frac{1+Fr_1}{2}\,q_{xW} + \frac{1-Fr_1}{2}\,q_{xP} - \frac{c_1}{2}\,(1-Fr_1^2)\,(h_P-h_W) \tag{5}$$

The above obtained values of h_R and q_{xR} allow the evaluation of $\Sigma_{x1}^*=\Sigma_{xR}$ through an average of the series expansions around the R-values along the characteristic directions:

$$\Sigma_{xR} = \frac{\Sigma_{xW}+\Sigma_{xP}}{2} + \left(\frac{\partial\Sigma}{\partial h}\right)_1\left(h_R - \frac{h_W+h_P}{2}\right) + \left(\frac{\partial\Sigma}{\partial q_x}\right)_1\left(q_{xR} - \frac{q_{xW}+q_{xP}}{2}\right)$$

With some algebra we get:

$$\Sigma_{x1}^* = \frac{1+Fr_1}{2}\,\Sigma_{xW} + \frac{1-Fr_1}{2}\,\Sigma_{xP} - \frac{c_1}{2}\,(1-Fr_1^2)\,(q_{xP}-q_{xW}) \tag{6}$$

The evaluation of $M_1^*=M_R$ follows the same procedure taking the series expansion around R:

$$M_R = \frac{M_W+M_P}{2} + \left(\frac{\partial M}{\partial v}\right)_1\left(v_R - \frac{v_W+v_P}{2}\right) + \left(\frac{\partial M}{\partial q_x}\right)_1\left(q_{xR} - \frac{q_{xW}+q_{xP}}{2}\right)$$

According to the upwind rule: $v_R = v_W$ if $u_1>0$, $v_R = v_P$ if $u_1<0$, then using (5) and simple algebra, we get the following (7):

$$M_1^* = M_W + \frac{v_1}{2}\,(1-Fr_1)\,[(q_{xP}-q_{xW}) - c_1\,(1+Fr_1)\,(h_P-h_W)] \quad ; \quad u_1>0$$

$$M_1^* = M_P - \frac{v_1}{2}\,(1+Fr_1)\,[(q_{xP}-q_{xW}) + c_1\,(1-Fr_1)\,(h_P-h_W)] \quad ; \quad u_1<0$$

In the supercritical case the same computations are done after setting M=N=W in Fig. 2b, that is, taking the direction W-R for both the 1D characteristics.

It is readly realized that the above given formulas account for both the subcritical and the supercritial case if the Froude number is limited within the values +1 and -1.

The same procedure is applied, with the obvious modifications, to the y direction to obtain the y-reach extrapolated fluxes there required.

Boundary conditions and dry bottom problem.

The boundary conditions are assigned following the characteristic method applied to each 'line' of the two 1D systems (2). Free slip conditions are assigned to the walls setting to zero the normal flow component, so that the wall lines in the direction of the wall itself behave like interior lines. At the inflow and outflow boundaries one condition (of excitation or impedence) is needed to meet with the compatibility condition beared by the outgoing characteristic in the case of subcritical flow. If the flow is supercritical two conditions are needed at the inflow and none at the outflow boundaries. Critical inflow can be assigned by specifying either a flow variable or the energy head; no extra information is required for the critical outflow.

The initial state of the system must be specified also. In the case of dry bed, all the flow variables are set to zero: to avoid numerical singularities the water depth h is always fixed to a small positive number: say 10^{-6} m.

To fulfill the stability condition the time step is selected as the minimum of the local values:

$$\delta t = \frac{\delta x\, \delta y}{\delta x\, \text{MAX}(|u+c|,|u-c|) + \delta y\, \text{MAX}(|v+c|,|v-c|)} \qquad (8)$$

where $\delta x\ \delta y$ are the 'cell' dimensions.

LABORATORY EXPERIMENT ON THE IDEALIZED SUDDEN LEVEE BREAKING AND FOLLOWING PLANE SUBMERSION

After the preliminary tests of the code with the 1D well known exact solutions [6] and the 2D literature numerical results [2], a laboratory experiment was decided to get a better feeling (and a more sound proof) of the model performance.

The scheme of the experimental setup is described in Fig. 3.

At the sudden gate opening, the reservoir water spreads over the transparent plane, and then outflows from the exit side into a collecting tank.

In order to visualize the characteristics of the flow field the water is coloured with methylene blue ink and the tank is lighted from below. By this way it is possible to follow the position of the wave front and to evaluate the water depth from the colour intensity.

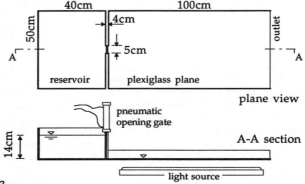

Fig. 3

The experiment was shot with a videocamera able to record 25 photograms per second, fixed on a scaffold 2m over the plane. Using a suitable software, it is possible to select from the whole film the photograms at given times after the water inflow.

The digitalized photograms are converted into black-and-white images, defined by a matrix relating each pixel to its level of grey from 0 (black) to 255 (white). These values are subsequently subdivided into 6 bands, so to obtain a series of contour lines as boundaries of selected colour ranges. In order to determine the relationship between water level and colour intensity, before every experiment some small trasparent basins with known water level were shot and their colour levels were used as landmark.

Two sets of experiments, starting from initial dry bed and initial water depth of 0.5 cm, are presented. For both the initial water level in the reservoir is of 14 cm.

The main flow characteristics in the two cases can be appreciated, for example, in Figg. 4a,4b relevant to the same time.

In the case of wet plane (Fig 4a) the water wave, originated at the inlet gate, first expands with a nearly semi-circular shaped belt. The front water levels are sensibly higher than both the initial still water and the rear high velocity flow levels. When the wave front reaches the lateral walls, by reflection two high level fluctuating wall regions are formed. They become thicker and thicker as long as the wave front travels towards the exit, till they join together at

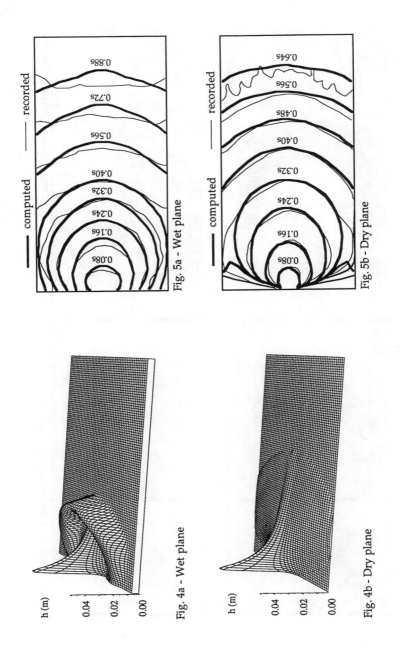

Fig. 5a - Wet plane

Fig. 5b - Dry plane

Fig. 4a - Wet plane

Fig. 4b - Dry plane

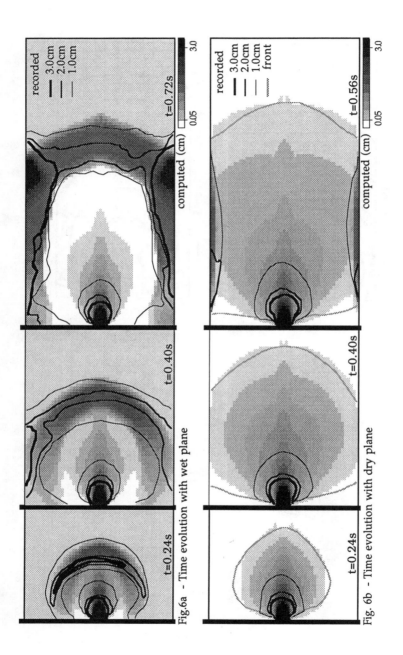

Fig.6a - Time evolution with wet plane

Fig. 6b - Time evolution with dry plane

the plane centerline and delimitate a triangle-shaped central region
of shallow fast flowing water.
When the dry bed is flooded, on the contrary, the wave travels
faster: it is characterized by continuous rising levels from the front
line up to the maximum at the gate entrance and the two wall
reflection regions do not join before the exit.
When the wave front arrives at about three quarters of the plane
we can observe the breaking up of the front line; only some
"fingers" of water advance quickly, while the remaining part of the
front slows down. Obviously from this instant the experimental
images are less meaningful for the comparison with the numerical
results.

The simulation of the observed flow fields has been performed on
a basic 50x100 square cell network with side of 1 cm: close to the
gate the mesh has been refined by halving the side of the cells.
The time step computed with the above formula (8) has been
reduced by .95 and a Manning coefficient of n=.01 has been
assumed. It has been verified that the use of a smooth wall formula
like Blasius' does not produce sensible difference in the results as
the flow is essentially controlled by the inertial effects and the
pressure forces.
In Figg. 5a, 5b several computed versus recorded front positions are
displayed at selected times after the opening, while the computed
water depths are compared with the observed ones in Figg. 6a, 6b.
There is a quite reasonable agreement between observation and
simulation in both the cases. The main depth distribution features
are retained and the front line celerity is in good agreement with
the observed one (at last before the 'fingering' in the dry bed
flooding occurs). The differences observed at the centerline and
close to the walls are probably due to grid effects and to the
connected numerical diffusion.
The local Froude number distribution shows that the flow
undergoes to critical transitions both in its acceleration through
the gate opening and when it is slowed down by the lateral wall
reflection and near the front line. The code proves to deal with
such transitions without troubles.

FAR FIELD SIMULATIONS

As above stated the mathematical description of the flow is based
on the shallow water equations (1) dropping out the convective
term that, with obvious simplification, read:

$$\frac{\partial h}{\partial t} + \frac{\partial q_x}{\partial x} + \frac{\partial q_y}{\partial y} = 0 \qquad (9a)$$

$$\frac{\partial u}{\partial t} + g\frac{\partial h}{\partial x} = g\,(S_{0x} - S_{fx}) \;\; ; \quad \frac{\partial v}{\partial t} + g\frac{\partial h}{\partial y} = g\,(S_{0y} - S_{fy}) \qquad (9b)$$

The differential problem is still of hyperbolic type: its proper position needs, therefore, the knowledge of initial and boundary conditions. The relaxation of the convective terms considerably simplifies the problem of proper defining the boundary conditions: since the supercritical flow cannot occur, a single condition is needed at every boundary point [5]. The value of the entering discharge or the water depth are proper conditions for the inflow boundaries, a functional relation between discharge and water depth is suitable for the outflow boundaries, while the no-flux condition is appropriate for the walls.

The space and time scales typical of the far field problem make the choice of an implicit scheme attracting: the numerical solution of system (9) is actually obtained by means of a finite difference fully implicit scheme working on a staggered grid space discretization.

The integration domain is approximated by a system of non-homogeneous rectangular contiguous cells over which the value of water depth and bottom level are referred to the centeroid P.

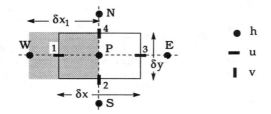

Fig. 7

Equation (9a) is first integrated over the cell to obtain, by means of the divergence theorem, the following discrete form :

$$\delta x \, \delta y \, [(h_P - h_{0P}) \,/\delta t = \delta y \, (q_{x1} - q_{x3}) + \delta x \, (q_{y2} - q_{y4}) \qquad (10)$$

The discharge flowing through the cell side between two nodes, according to equations (9b), is driven by the depth gradient and by the corresponding sources. The dynamic equation is essentially monodimensional if a proper linearization of the Manning formula is assumed: $S_{fx} = \lambda u$; the integration of (9b) over the shaded cell of Fig. 7 brings to the following expression of q_{x1} as a function of the unknowns h_W and h_P :

$$q_{x1} = u_1 h_1 \; ; \quad h_1 = (h_P + h_W)/2 \; ; \quad u_1 = \alpha - \beta (h_P - h_W)$$

$$\alpha = \frac{u_0 + g \, \delta t \, S_{0x}}{1 + g \, \delta t \, \lambda} \; ; \quad \beta = \frac{\delta t}{\delta x_1 \, (1 + g \, \delta t \, \lambda)}$$

The above unit discharge expression is replaced by a suitable function of the water depths across the side of the cell when a local flow control structure has to be simulated.

Since the side discharges are non linear functions of the water depth in P and in its four neighbouring cells, eqn. (10) can formally be written as:

$$F_P (h_P, h_W, h_S, h_E, h_N) = \delta x \, \delta y \, [(h_P - h_{0P})/\delta t - \delta y(q_{x1} - q_{x3}) - \delta x(q_{y2} - q_{y4}) = 0$$

Looping all over the cells and properly incorporating the boundary conditions, a system of non linear equations as the cell unknown water depths is written. The solution of such a system is obtained by means of a non linear Gauss-Seidel iteration technique with local updating of the coefficients to account for the non linearity of the problem. For the details reference is made to [7].

In Figg. 8a,8b the 1987 Ardenno (North of Italy) plane flooding, computed 3 and 6 1/2 hours after the breaking , are reported as an example.

The code has been employed to get a large scale simulation of the 1966 Florence flooding. Such an attempt has been suggested by the idea that the flow field, seen at a sufficiently large scale, is not topologically too different from the ground water flow occurring in a fracturated medium. If one thinks that a computational cell typically contains several roads, buildings, squares, etc., so that the water can reach a point inside entering from every side , and that the time scales are large enough to make it possible, the field can be schematized as a 2-D continuum introducing a suitable field parameter that could be seen as an 'urban porosity'.

Obviously a detailed description of the flow at the scale of the road geometry needs a channel network topological schematization rather than the 2-D continuous one. Such a point of view, in fact, has been followed to detail the simulation in the town hystorical center [8], matching the necessary boundary conditions with the results of the 2-D simulation.

As an example two pictures of 1966 Florence flooding simulation taken from [8] are reported in Figg. 9a,9b. The pictures refer to the situation at 9 a.m. and to the maximum flooding occurred at about

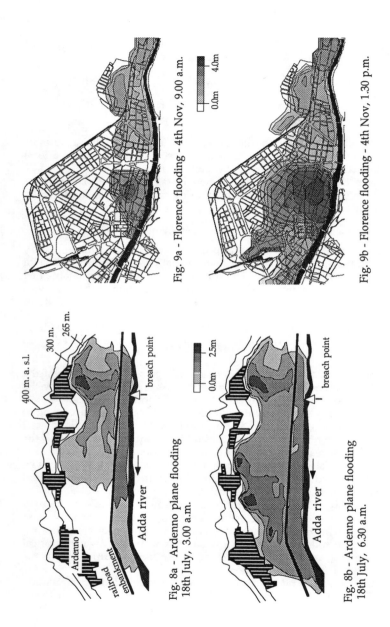

Fig. 9a - Florence flooding - 4th Nov, 9.00 a.m.

0.0m   4.0m

Fig. 9b - Florence flooding - 4th Nov, 1.30 p.m.

400 m. a. s.l.

300 m.

265 m.

Ardenno

railroad
embankment

breach point

Adda river

Fig. 8a - Ardenno plane flooding
18th July, 3.00 a.m.

0.0m   2.5m

breach point

Adda river

Fig. 8b - Ardenno plane flooding
18th July, 6.30 a.m.

1: 30 p.m. of Nov 4th. They compare fairly well with the reported observations.

FINAL REMARKS

In large scale descriptions the numerical simulation has proved to be a valuable tool to get rational scenarios of plain flooding relevant to possible point source hypotheses to be used in the frame of disaster prevention engineering, despite the characteristics of the real phenomena cannot be foreseen due to their intrinsic complexity and because of the influence of unpredictable causes.
In near field simulations, where the convection effects make the mathematical problem heavier, the rather simple (and robust) first order scheme described has proved capable to depict the main flow characteristics of a laboratory smooth plane submersion.
Maybe 2nd order modelling could lead to even better results for the problem, but the question can be raised whether the real world source effects, whose strenght is badly defined and modelled, are so important to vanify the advantages of refined (i.e. 2nd order) modelling now popular in gasdymamics applications.

BIBLIOGRAPHY

[1] Hirsch C., "Numerical computation of internal and external flows", John Wiley & Sons, 1990
[2] Alcrudo F., Garcia-Navarro P., "A TVD scheme in finite volumes for the simulation of 2D discontinuous flows", Jornadas de encuentro trilateral para el estudio de la hidraulica de las ondas de submersion, Zaragoza, 1992
[3] Cunge J.A., Holly F.M., Verwey A., "Practical aspects of computational river hydraulics", Pitman Adv. Pubbl. Comp., 1980
[4] Abbott M.B., Basco D.R., "Computational fluid dynamics", Longman Scientific and Technical, 1989
[5] Daubert A., Graffe O., "Quelques aspects des écoulements presque horizontaux à deux dimensions en plan et non permanents", La Houille Blanche, 1967
[6] Braschi G., Gallati M., "A conservative flux prediction algorithm for the explicit computation of transcritical flow in natural streams", Hydrosoft92, Valencia 1992
[7] Braschi G., Gallati M., "Simulation of a levee-breaking submersion of planes and urban areas", HYDROCOMP89, Elsevier Appl. sc., 1989
[8] Braschi G., Gallati M., Natale L., "La simulazione delle inondazioni in ambiente urbano", CNR-GNDCI, 1990

2D FINITE ELEMENT MODELLING OF FLOODING DUE TO RIVER BANK COLLAPSE.

L. D'Alpaos, A. Defina
Institute of Hydraulics "G. Poleni", University of Padua
Via Loredan, 20 - 35131 Padua ITALY

B. Matticchio
C.N.R. I.S.D.G.M.
S. Polo 1364 - 30125 Venice ITALY

ABSTRACT

To compute the flood propagation over an initially dry area a new set of 2D equations has been recently proposed which accounts for the ground unevenness effects.
The paper presents a 2D mathematical model which solves these equations by a staggered finite element scheme. The whole domain is divided into triangular elements. Water elevations are considered linearly varying between element nodes while velocities are considered constant over the element.
The proposed model is applied to study the flood propagation along the Brenta river considering, in addition, a possible river bank collapse during the high stage.
The influence of the threshold depth Y_{lim} upon the celerity of the front propagation over the dry area behind the broken bank is also investigated with the model.

INTRODUCTION

In the application of two-dimensional mathematical models to the prediction of velocities and water depths, the ground surface is usually assumed to be smooth. The effects of terrain unevenness and the presence of ditches which can make the flooding front propagation easier and increase its celerity are thus neglected.

Based on the assumption that usually in very shallow flows dissipations are largely predominant over inertial effects (certainly this is not the case of 1D dam break problem for a smooth bed) a new set of 2D equations was recently proposed by the authors [1].

The new equations allow a very realistic description of the small water depth flows like flood propagation over an initially dry area or overland flow in a catchment during a rainfall event.

Two parameters, η and H, are introduced to modify the well known De S. Venant equations both depending upon a characteristic length Y_{lim} which is strictly related to the ground unevenness.

The effectiveness of the new model, postulated by the authors, is checked through an application to a field case.

THE MATHEMATICAL MODEL

The model solves the vertically integrated momentum and continuity equations for shallow free surface flow proposed by Defina e al. [1] :

$$\frac{\partial h}{\partial x} + \frac{1}{g}\frac{du}{dt} + \frac{q_x \cdot |\mathbf{q}|}{K_s^2 \cdot H^{10/3}} = 0 \tag{1}$$

$$\frac{\partial h}{\partial y} + \frac{1}{g}\frac{dv}{dt} + \frac{q_y \cdot |\mathbf{q}|}{K_s^2 \cdot H^{10/3}} = 0 \tag{2}$$

$$\eta \cdot \frac{\partial h}{\partial t} + \nabla \mathbf{q} = w \tag{3}$$

where h is the free surface elevation, u, v are the depth-averaged velocity in x, y directions, g is the gravitational acceleration, K_s is the Strickler friction coefficient, t is time, $\mathbf{q}$ is the depth integrated velocity with components q_x and q_y and w is a source/sink term. The local percentage of wetted domain h and the equivalent depth H are given as:

$$\eta = \begin{cases} e^{-0.7\left(1-\frac{Y_a}{Y_{lim}}\right)^2} & \text{if } Y_a < Y_{lim} \\ 1 & \text{if } Y_a > Y_{lim} \end{cases} \quad (4)$$

$$\frac{H}{Y_{lim}} = \frac{1}{4}\left\{\frac{Y_a}{Y_{lim}} + \sqrt{\left(\frac{Y_a}{Y_{lim}}\right)^2 + 0.5}\right\}\cdot\left[1+\tanh\left(\frac{2\cdot Y_a}{Y_{lim}}+2\right)\right] \quad (5)$$

where $Y_a = h - \overline{h_f}$, $\overline{h_f}$ is the local mean bottom elevation and Y_{lim} is the largest amplitude of ground ripples.

Convective accelerations have been neglected in momentum equations (1) and (2) as dissipations are largely predominant during slow flooding.

Equations (1) to (5) are discretized in time by using a centered finite difference approximation.

Assuming:

$$\psi = \left[\frac{1}{gH'\Delta t} + \frac{|\mathbf{q'}|}{K_s^2 \cdot H'^{10/3}}\right]^{-1} \quad (6)$$

$$\varphi_x = -\frac{q'_x}{gH'\Delta t} \quad , \quad \varphi_y = -\frac{q'_y}{gH'\Delta t} \quad (6')$$

equations (1) and (2) take the form:

$$q_x = -\psi\cdot\left(\frac{\partial h}{\partial x}+\varphi_x\right) \quad (7)$$

$$q_y = -\psi\cdot\left(\frac{\partial h}{\partial y}+\varphi_y\right) \quad (8)$$

where the prime indicates values at the previous time level.
Substitution of equations (7) and (8) into equation (3) leads to

$$\eta\cdot\frac{\partial h}{\partial t}-\nabla\cdot\left[\psi\cdot(\nabla h + \vec{\varphi})\right] = w \quad (9)$$

Equations (7) to (9) are solved by a standard semi-implicit finite element method.
The two dimensional domain is divided into triangular

elements. Water levels are considered linearly varying between element nodes, while depth-integrated velocities are considered constant over the element.
The main unknown h is approximated by

$$\hat{h} = \sum_{n=1}^{N} h_n \, \xi_n$$

where ξ_n is the linear shape function and N is the total number of nodes.
Application of the Galerkin's method and the Green's first identity to equation (9) gives

$$\sum_e \eta^e \cdot \int_{A^e} \frac{h_n - h'_n}{\Delta t} \cdot \xi_n^e \cdot \xi_m^e \cdot dA^e + Q_m +$$
$$+ \sum_e \int_{A^e} \psi^e \cdot \left[\frac{h_n + h'_n}{2} \nabla \xi_n^e + \vec{\phi}^e \right] \cdot \nabla \xi_m^e \cdot dA^e = 0 \qquad (10)$$

where repeated subscripts imply summation, A^e is the generic element area, Q_m is the total node outflow and

$$\eta^e = \frac{1}{A^e} \int_{A_e} \eta \cdot dA_e$$

is assumed constant during each time step.
Once equation (10) is solved for the surface elevations h_n at each node then the depth-integrated velocities are easily computed by backward substitution from equations (7) and (8).

MODEL APPLICATION

The mathematical model was applied to a field case: the flood propagation along the river Brenta (Italy) considering also the hypothetical case of a breach formed in the river bank.
The studied river reach, which extends along a distance of about 10 km from Bassano del Grappa to Tezze, is described using a mesh of 1365 triangular elements and 803 nodes as shown in Fig. 1.
Two different values of the Strickler coefficient were used: $K_s=25$ $m^{1/3}s^{-1}$ for the main channel and $K_s=10$ $m^{1/3}s^{-1}$ elsewhere. A constant value, $Y_{lim}=0.3$ m, was assumed in the whole domain.

Two types of conditions were specified at the open boundaries: discharge as a function of time at the upstream end of the river and a stage-discharge relationship at the downstream open boundaries.

Steady state flow at a rate of 50 m^3/s was assumed as initial condition (see Fig. 1): the flow, in this case, occurs only in the main channel while most of the domain is completely dry.

A first simulation was run in which the measured discharges of the november 1966 flood event were prescribed and no bank break was considered.

Results of this simulation are depicted in Fig. 2 to Fig. 3. No detectable instabilities were seen in computed values of free surface elevation and flow velocity.

Fig. 2 is a plan view of the water depths at three different times during the flood propagation. Most of the flood plains are gradually wetted but there remain some dry islands even during the peak flow (t=30^h). The water depth representation results in a stepwise fashion because of the staggered scheme used: ground elevations are prescribed at the element centre while free surface elevations are computed at the element nodes.

The prescribed hydrograph and the computed discharge at the downstream end are plotted in Fig. 3.

Only a very small reduction of the peak discharge can be detected due to the very steep river bed (average bottom slope $\approx 4.5\%$₀).

Based on the peak lag, a celerity c_Q of about 5 m/s is estimated which is only 0.7 times the largest flow velocity.

In the second run the collapse of the river bank near Rivarotta (See Fig. 1) at time t=27^h was roughly simulated by lowering about 60 m of the bank top to the local ground elevation linearly in 10 minutes.

Some results of this simulation are shown in Fig. 3 to Fig. 5.

In Fig. 3 the computed downstream hydrograph clearly shows the effects of the outflow through the breach. At t=27^h a small but sharp decrease of discharge is evident and the peak is reduced to about 90%.

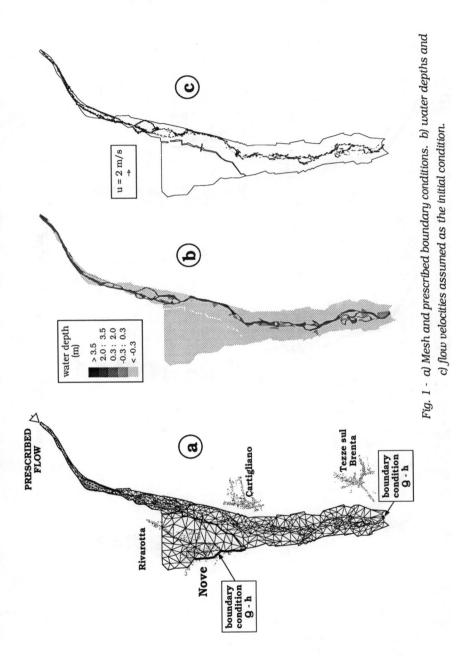

Fig. 1 - a) Mesh and prescribed boundary conditions. b) water depths and c) flow velocities assumed as the initial condition.

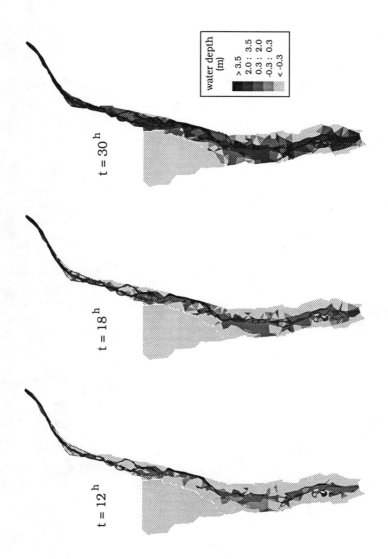

Fig. 2 - Water depths at diffent times during the flood propagation.

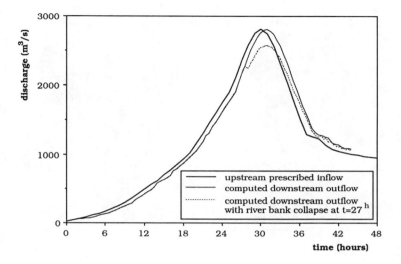

Fig. 3 - *Prescribed upstream discharge (november 1966 flood event) and computed downstream discharge without bank break (full line) and considering bank break at t=27^h (dashed line)*

Flood propagation over the initially dry area including the city of Nove is depicted in Fig. 4 showing plan views of both water depths and velocity fields. The flooding front inside this area propagates slowly in the direction of the mean ground slope and after reaching the main road at the left side of the domain it overflows out.

A rough estimate of the front propagation celerity is possible if plotting the time spent to reach some nodes lying along the mean front path as shown in Fig. 5. Soon after the initial fast advance, the front slows down and propagates at a velocity of about 0.35 m/s.

Finally, the influence of Y_{lim} was investigated by running the model assuming for the flooded area two different values of this parameter: Y_{lim} = 0.6 m and Y_{lim} = 1 cm.

In Fig. 5 the wetting front position is plotted versus time along the main front path. Doubling the value of Y_{lim} (from 0.3 m to 0.6 m) results in an average celerity increase from 0.5 m/s to 2.8 m/s, that is more then five times. The effect of changing

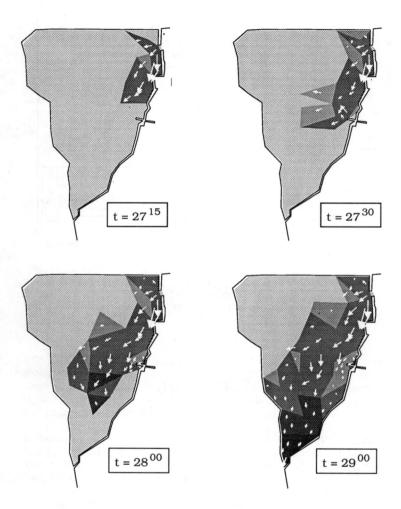

Fig. 4 - Water depths at different times during the front propagation over the initially dry area behind the river Brenta right bank soon after its collapse. The water depth scale is the same of Fig. 2. The velocity scale is the same of Fig. 6.

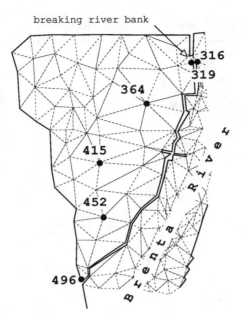

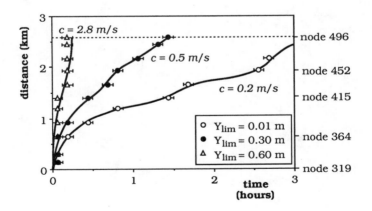

*Fig. 5 - Above : enlarged view of the mesh covering the flooded area.
Below : distance covered by the flooding front along its mean
path versus time for three different values of Y_{lim} .*

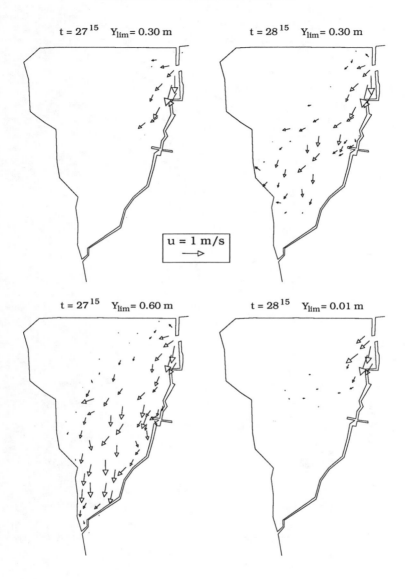

$t = 27^{15}$ $Y_{lim} = 0.30$ m $t = 28^{15}$ $Y_{lim} = 0.30$ m

$u = 1$ m/s

$t = 27^{15}$ $Y_{lim} = 0.60$ m $t = 28^{15}$ $Y_{lim} = 0.01$ m

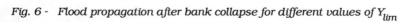

Fig. 6 - Flood propagation after bank collapse for different values of Y_{lim}

Y_{lim} is also evident in Fig. 6 where the velocity fields are plotted at $t = 27^{15}$.

This result highlights the importance of Y_{lim} and the need for an accurate estimate of its value.

The case $Y_{lim} = 1$ cm, which is not realistic, was run because most of the models presented in the literature and based upon the classic De S. Venant equations, consider a sharp transition from dry to wet conditions when the local free surface elevation exceeds a threshold value which is usually very small. The average front celerity in this case is strongly reduced.

The velocity fields at $t = 28^{15}$ plotted in Fig. 6 clearly show the reduced advance velocity of the front when $Y_{lim} = 1$ cm.

CONCLUSIONS

A 2D mathematical model for very shallow water flows based upon a new set of equations was developed and run to simulate flood propagation along a river with large, initially dry, flood plains. A localized river bank collapse was also simulated to analyze the front propagation during the flooding of a small area. The computed results showed a satisfactory stability of the scheme also during wetting and drying phenomena.

The postulated importance of accounting for ground unevenness and thus of Y_{lim} was also confirmed.

REFERENCES

[1]   Defina, A. D'Alpaos, L. Matticchio, B. *A New Set of Equations for Very Shallow Water and Partially Dry Areas Suitable to 2D Numerical Models.* This conference

A NEW SET OF EQUATIONS FOR VERY SHALLOW WATER AND PARTIALLY DRY AREAS SUITABLE TO 2D NUMERICAL MODELS

A. Defina, L. D'Alpaos
Institute of Hydraulics "G. Poleni", University of Padua
Via Loredan, 20 - 35131 Padua, ITALY

B. Matticchio
C.N.R. I.S.D.G.M.
S. Polo 1364 - 30125 Venice, ITALY

ABSTRACT

When a 2D model is used to compute the flood propagation over an initially dry area some difficulties arise due to the changes in the computational domain as large parts of it are flooded and dried during the phenomenon.
A second problem concerns the representation of soil surface: usually it is described as a piecewise constant or a piecewise linearly varying surface hence neglecting the ground unevenness. For very small water depths, unevenness plays instead a very important role.
In this paper, new dynamic and mass balance equations for 2D free surface flows are presented which, by means of a conceptual sub-model, account for the ground unevenness effects and allow a very realistic description of flooding and drying phenomena.
It is also shown that for a very smooth ground surface or for relatively large water depths, the new equations reduce to the well known De S. Venant equations.

INTRODUCTION

In the application of two-dimensional mathematical models to the prediction of velocities and free surface elevations, the bottom is usually described as a smooth surface, thus neglecting terrain unevenness: ground elevations are assumed piecewise constant or piecewise linearly varying inside the computational cells.

Such a simplification can be undertaken only for water depths which are large when compared with ground macro roughness but it is surely not effective for the thin sheet of water flowing over an initially dry surface during flooding phenomena.

This loss of effectiveness, as it will be shown, must be ascribed to a worst estimation of both friction and storage. This is because numerical models usually assume a sharp transition, often controlled by a threshold water depth, from wet to dry conditions inside the generic computational cell [1], [2], [3], [4].

Such a transition, in fact, occurs instantaneously only at a single point, that is for an infinitesimal portion of the domain. Over a small but finite area (e.g. a computational element), instead, transition from wet to dry conditions or vice versa, takes place gradually due to the ground surface irregularities.

THE PROPOSED NEW SET OF EQUATIONS

To take account of the ground unevenness in the mass balance equation Defina and Zovatto [5] proposed a new interesting approach which is here briefly recalled and refined.

The conceptual model adopted to simulate flooding or drying processes is based on the physics of the phenomenon as previously described and refers to a small but finite portion Ω of the domain.

Let $\overline{h_f}$ be the mean bottom height in Ω. We introduce the threshold depth Y_{lim} as the difference between the highest ground elevation and $\overline{h_f}$, in Ω.

When the free surface elevation h is lowered to $\overline{h_f} + Y_{lim}$ then the flow domain begins to dry and some portions of the ground gradually emerge. We can assume that completely dry conditions are achieved only when h becomes lower than about $\overline{h_f} - Y_{lim}$ as shown in Fig.1.

It is convenient to let η be the wetted percentage of Ω as a function of h.

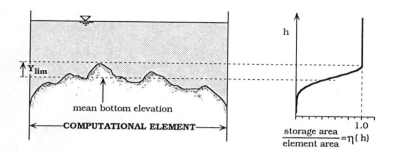

Fig. 1 -   *percentage of wetted area η as a function of
free surface elevation.*

The new expression for continuity is then written as:

$$\eta \cdot \frac{\partial h}{\partial t} + \nabla \mathbf{q} = w \qquad (1)$$

where $\mathbf{q}$ is the depth integrated velocity and w is a
source/sink term.

Equation (1) is an extension of the usual 2D fluid continuity
equation: if we let $\Omega \rightarrow 0$ then we find that $Y_{lim} \rightarrow 0$ and η
approaches the Unit step function. Hence when equation (1) is
applied only to the wet part of the domain then $\eta=1$.

More properly η may be interpreted as a h dependent
"storativity coefficient" analogous to the one used in
groundwater hydraulics.

An analytical expression for η was found by examining a
number of real topographical profiles: assuming $Y_a = h - \overline{h_f}$, the
dependence of η upon Y_a/Y_{lim} can conveniently be written in
the form

$$\eta = \begin{cases} e^{-\alpha \left(1 - \frac{Y_a}{Y_{lim}}\right)^2} & \text{if} \quad Y_a < Y_{lim} \\ 1 & \text{if} \quad Y_a > Y_{lim} \end{cases} \qquad (2)$$

where $\alpha=0.7$.

The function η given by equation (2) is plotted in Fig. 2.

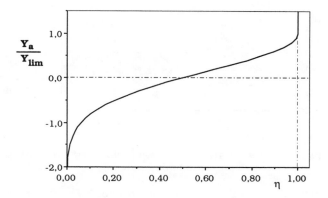

Fig. 2 - Percentage of wetted area η as a function of Y_a/Y_{lim}.

The ground unevenness also affects the dynamics of the flow. For very small water depths energy losses dominate, thus the attention is focused on the dissipation term in the dynamic equations.

We assume, first, that ground "macro roughness" does not produce minor losses. We then consider a small square sub-domain $\Omega = L \times L$ with a constant cross section as depicted in Fig. 3.

For a steady uniform flow we may write:

$$q = \frac{Q}{L} = \frac{K_s \cdot \sqrt{J}}{L} \int_L Y(h,\xi)^{5/3} \cdot d\xi \qquad (3)$$

where Q is the discharge, L is the cross section width, J is the mean energy slope and K_s is the Strickler coefficient.

In equation(3) the water depth Y is defined as follows:

$$Y(h,\xi) = \begin{cases} h - h_f(\xi) & \text{if } h \geq h_f(\xi) \\ 0 & \text{if } h < h_f(\xi) \end{cases} \qquad (4)$$

where h_f is the bottom elevation. In writing equation (3), we have neglected any transverse dynamic interaction.

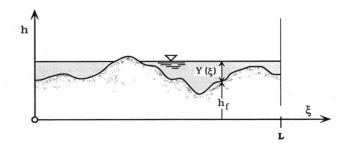

Fig. 3 - Schematic representation of a cross section of a 2D flow domain.

We introduce now the following expression to evaluate energy losses per unit length:

$$J = \frac{q^2}{K_s^2 \cdot H^{10/3}} \tag{5}$$

where H is a characteristic depth.
Combining equations (3) and (5) allows to write:

$$H = \left[\frac{1}{L} \int_L Y(h,\xi)^{5/3} \cdot d\xi \right]^{3/5} \tag{6}$$

This result, together with equation (4), after some algebra leads to

$$\frac{H}{Y_{lim}} = \left[\int_{-\frac{h_f - h_{f_{min}}}{Y_{lim}}}^{min\{Y_a/Y_{lim},1\}} \left(\frac{Y_a}{Y_{lim}} - \varsigma \right)^{5/3} \cdot \frac{\partial \eta}{\partial \varsigma} \cdot d\varsigma \right]^{3/5} \tag{7}$$

where $\zeta = (h_f - \overline{h}_f) / Y_{lim}$ and $h_{f_{min}}$ is the lowest bottom elevation.

As $\partial \eta / \partial \zeta = 0$ for $h_f < h_{f_{min}}$ then, from equation (7), H/Y_{lim} only depends on Y_a/Y_{lim}. Moreover, when $\Omega \to 0$ then $Y \equiv Y_a$ and equation (6) gives $H = Y_a$.

Equation (7), with η given by equation (2), was integrated

- 76 -

numerically and the result is plotted in Fig.4.
A suitable analytical approximation for H is given by:

$$\frac{H(h)}{Y_{lim}} = \frac{1}{2}\left\{\frac{Y_a}{Y_{lim}} + \sqrt{\left(\frac{Y_a}{Y_{lim}}\right)^2 + 0.5}\right\} \cdot Z\left(\frac{Y_a}{Y_{lim}}\right) \qquad (8)$$

where

$$Z\left(\frac{Y_a}{Y_{lim}}\right) = \frac{1}{2}\left[1 + tgh\left(\frac{2 \cdot Y_a}{Y_{lim}} + 2\right)\right] \qquad (8')$$

which is also plotted in Fig.4.

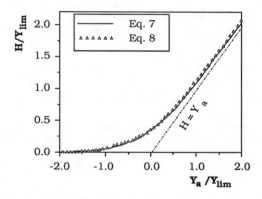

Fig. 4 -   H/ Y_{lim} as a function of Y_a/Y_{lim}.

Ground unevenness also affects local and convective acceleration. Suitable Coriolis coefficients should be introduced in the dynamic equation terms but as dissipations largely dominate over inertia then these corrections have not been considered in the present analysis.
Consequently, for a shallow two dimensional flow, dynamic and continuity equations can be finally written as:

$$\frac{\partial h}{\partial x} + \frac{1}{g}\frac{du}{dt} + \frac{q_x \cdot |\mathbf{q}|}{K_s^2 \cdot H^{10/3}} = 0 \qquad (9)$$

$$\frac{\partial h}{\partial y} + \frac{1}{g}\frac{dv}{dt} + \frac{q_y \cdot |\mathbf{q}|}{K_s^2 \cdot H^{10/3}} = 0 \qquad (10)$$

$$\frac{\partial h \cdot \eta}{\partial t} + \nabla \mathbf{q} = w \qquad (11)$$

where u, v are the components of the flow velocity.

We have seen previously that (9), (10) and (11) are consistent with De S. Venant equations. Moreover the usual extensions over a discrete domain of De S. Venant equations are also effective for large water depths, as when $h \gg Y_{lim}$ then $\eta = 1$ (see Fig. 2) and $H \cong Y_a$ (see Fig. 4).

When water depth is instead small in comparison with ground unevenness then equations (9), (10) and (11) seem to be much more effective in numerical models as they take explicitly account, though synthetically, of the ground unevenness through η and H which both depend upon Y_{lim}.

Certainly Y_{lim} must be evaluated with accuracy.

When $\Omega \to 0$, we have seen, $Y_{lim} \to 0$; on the other hand, for a finite portion of domain, Y_{lim} is greater than zero. We can therefore expect that Y_{lim} depends upon both the ground roughness and the computational elements area.

A second question regards the effectiveness of introducing such a parameter as a measure of ground unevenness for practical applications.

To answer these questions it is convenient to consider the function η from a statistical point of view: η may be considered as the probability P that bottom height inside Ω does not exceed h:

$$\eta \left(Y_a / Y_{lim} \right) = P(h_f < h) \qquad (12)$$

Therefore $\partial \eta / \partial (Y_a / Y_{lim})$ is the distribution of bottom elevations inside Ω and an effective measure of ground unevenness is then given by the Standard Deviation σ_{h_f} of bottom elevations. In addition we can assume that

$$Y_{lim} \cong 2\sigma_{h_f} \qquad (13)$$

which relates the statistics of the ground surface to Y_{lim}.

For a ground profile (in two dimensions), when plotting

Standard Deviation σ_{h_f} versus the length Λ of a generic sub-domain in a log-log chart it should result in a straight line due to the *self-affine* nature of terrain elevations. The slope $\mathcal{H}$ of such a line is the Hausdorff measure and it is related to the local fractal dimension D_L as : $D_L = 2 - \mathcal{H}$ [6], [7].
It then follows that:

$$Y_{lim} \propto \Lambda^{\mathcal{H}} \qquad (14)$$

It is widely accepted that natural ground profiles bear a striking resemblance to the traces of *fBm* (fractional Brownian motion) with $\mathcal{H} \approx 0.5$. Equation (14), in this case, holds also in three dimensions [7].
Recently Matsushita and Ouchi [6] suggested that a real topographical surface is characterized by, at least, two regimes: local and global. For global structures $\mathcal{H}$ does not differ much from 0.5 while for local features, because of diffusion effects due to small scale erosions, usual values for $\mathcal{H}$ are larger then 0.5 and are often close to unity.
Fig. 5 shows an example of mountain profile crossing the River Meschio valley (TV, Italy). The Standard Deviation of ground altitude versus the length Λ of a generic sub-domain is plotted in Fig. 6 and clearly shows two distinct regimes: when the length Λ is greater than about 1000 m then $\mathcal{H} \approx 0.5$ while for the local regime it is found $\mathcal{H} \approx 1$.
For a 2D sub-domain, as $\Omega \approx \Lambda^2$, then equation (14) becomes

$$Y_{lim} \propto \Omega^{\mathcal{H}/2} \qquad (15)$$

As we are not interested in the very small ground ripples, which are considered as hydraulic roughness, we can conclude that, as a first approximation, a linear variation of Y_{lim} with the square root of elements area Ω should be assumed. It can also be shown that when a piecewise linearly varying surface is assumed instead, then a quadratic dependence of Y_{lim} from Ω is to be expected.

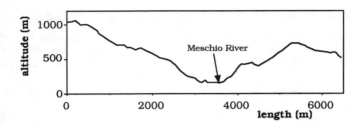

Fig. 5 - *Topographical profile near Vittorio Veneto (TV). The altitude variation is doubly exaggerated*

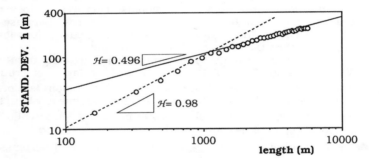

Fig. 6 - *Standard Deviation of altitude versus the length between many pairs of points on the curve of Fig. 5.*

CONCLUSIONS

New dynamic and mass balance equations for 2D free surface flows are proposed which, by means of a conceptual sub-model, take account of the ground unevenness effects and allow a very realistic description of the small water depth flows (e.g. flood propagation over an initially dry area or overland flow in a catchment during a rainfall event).

Two parameters, η and H, are introduced to modify the well known De S. Venant equations both depending upon a characteristic length Y_{lim} which is strictly related to the ground unevenness. Moreover, when the present equations are used in a numerical model, due to the finite dimension of computational elements, a dependence of Y_{lim} upon the element area must be accounted. This is related to the discrete approximations adopted to describe the ground surface, and a linear dependence can be expected if ground elevations are assumed piecewise constant over the computational cells. The effectiveness of the new set of flow equations strictly depends on the possibility to obtain an accurate estimation of Y_{lim}. Additional research should hence be undertaken in order to develop numerical techniques for estimating this parameter.

REFERENCES

[1] Stelling, G. S. Wiersma, A. K. Willemse, B. T. M. *Practical aspects of accurate tidal computations.* J. Hydraulic Engng. Am. Soc. Civ. Engrs, 1986, 112, No 9, Sept., 802-817.

[2] Falconer, R. A. Owens, P. H. *Numerical simulation of flooding and drying in a depth-averaged tidal flow model.* Proc. Instn Civ. Engrs, Part 2, 1987, 83, Mar., 161-180.

[3] Falconer, R. A. Chen, Y. *An improved representation of flooding and drying and wind stress effects in a two-dimensional tidal numerical model.* Proc. Instn Civ. Engrs, Part 2, 1991, 91, Dec., 659-678.

[4] Natale, A. Savi, S. *Espansione di onde di sommersione su terreno inizialmente asciutto,* 1991, Idrotecnica, No 6, Nov.-Dec, 397-406.

[5] Defina, A. Zovatto, L. *Modellazione matematica delle zone soggette a periodico prosciugamento in un bacino a marea.* Istituto Veneto di SS.LL.AA., Rapporti e Studi. Volume XII, 1994 (in press)

[6] Matsushita, M. Ouchi, S. *On the self-affinity of various curves.* Fractals in Physics - Essays in honour of Benoit B. Mandelbrot. Proceedings of the Int. Conf. honouring Benoit B. Mandelbrot on his 65th birthday. Vence, France, 1-4 October, 1989.

[7] Turcotte, D. L. *Fractals and Chaos in geology and geophysics,* Cambridge University Press 1992.

A CONTROL VOLUME FINITE ELEMENT METHOD FOR THE SOLUTION OF 2-D OVERLAND FLOW PROBLEMS

P. Di Giammarco, E. Todini

(Hydraulic Engineering Institute, Bologna University, Viale Risorgimento 2, 40136 Bologna, Italy).

ABSTRACT

This article introduces the Control Volume Finite Element Method (CVFEM) for the solution of 2-D overland flow problems. The CVFEM integrates the advantages of both Finite Element (FE) and Integrated Finite Difference (IFD) methods.

More in detail the advantages of the CVFEM can be summarised as a better physical interpretation of results as well as a reduction in the unknowns of the problem to be solved combined with a more effective representation of the gradients which is particularly important in overland flow problems.

In this paper the derivation of the algorithm is presented and a number of numerical problems are discussed.

After the description of the proposed method, a test is described which was carried out in order to verify the obtained results with respect to a theoretical solution.

1. INTRODUCTION

Two-dimensional overland flow problems were initially treated via Finite Difference (FD) methods, subsequently, after the development of the Integrated Finite Difference Method (IFDM) and of the Finite Element method (FEM) in porous media flow problems, these latter numerical techniques have also been applied to overland flow problems.

Difficulties still remain in the solution by the mentioned schemes as reported in the literature: each method has a number of pro and cons. With regard to the study of flood events in two-dimensional domains, although the problem to be resolved is similar to that of flow in porous media (inasmuch as various authors demonstrate how a parabolic type problem can be derived from the original hyperbolic problem), there are however aspects, such as the greater magnitude of the gradients and the greater velocities reached, which pose enormous problems for the traditional finite element method in which the number of unknowns - the heads - is taken to be equal to the number of nodes in the grid.

Therefore, in the most common applications, methods based on state vectors are used which present explicitly not just the heads at the nodes but also the components of the velocity vector, in two right-angled directions taken as reference, for each of the elements identified (Katopodes, [21]). In practical terms this requirement trebles the number of unknowns in the system of equations to be resolved when one wishes to use an implicit solution method. Many solutions proposed in the literature use not so much an implicit time integration method in the true sense but rather explicit methods of the conventional or forecast-correction type (Bechteler et al, [7]).

However, it should be stressed that implicit integration methods can only be avoided if extremely small time steps are used, bearing in mind the spatial dimensions of the discretisation meshes generally used and the propagation celerity of the perturbations involved, given the more or less marked steepness of the gradients and, especially, the limited height of the water levels which may even be null when starting from dry soil.

More recently in heat transfer and porous media flow (Patankar, [26]; Fung et al. [16]), the control volume finite element method (CVFEM) has been introduced; in addition to furnishing a more accurate description of gradients compared with the integrated finite difference method, this technique also permits the local balance of the scalar quantities such as mass and energy.

The introduction of the control volume finite element method not only permits a physical interpretation of the quantities that describe the flow terms (an aspect which is not altogether clear in the traditional finite element method), it also allows the problem to be addressed by considering just the heads at the nodes as unknowns, thus permitting rapidly converging implicit schemes to be developed.

The CVFEM is locally conservative (like the integrated finite difference method) and permits an accurate description of complex geometries and also of pronounced heterogeneities and anisotropies (like the finite element method).

Also presented is a numerical example which has been chosen not so much to highlight the characteristics peculiar to the CVFEM as to verify the calibration of the programme developed. In the example given, in fact, a comparison of the solutions obtained using the various methods shows how little they differ from each other and from the available theoretical solution.

2. THE METHODOLOGICAL APPROACH

The Control-Volume Finite Element Method

The CVFE method is based on assigning the conservation of mass, momentum and energy to each control volume, the shape and position of which is known with precision, as in the integrated finite difference method (Narasimhan et al., [23]; Hromadka et al, [19]; Todini et al. [32] and [33]). In this case attention is focused on a defined volume, fixed in space - the control volume - whose shape and position are chosen arbitrarily but are invariable, and the balance equations are written for each volume.

The method usually adopted for the integrated finite differences may be used for the spatial discretisation of the domain, though in this case the triangulation is preserved and forms the basis for the calculation of the gradients. The flexibility of the triangular element discretisation structure permits the accurate spatial reproduction of the domain, even in the case of complex geometries. The dual definition of the polygons (control volumes) which constitute the basis for determining the exchanges between adjoining nodes (and on which the momentum balance equations are set) ensures an extremely immediate physical interpretation of the quantities present in the discretisation equation. The validity of the mass, energy and momentum balances on each control volume is also assured. Further details on the CVFEM can be found in Patankar [26], while an example of application to the simulation of flow in porous media has been developed by Fung et al. [16].

<u>Description of the method.</u>   The control volume finite element method can be seen as a particular version of the weighted residues method used in the FE method (see Zienkiwics [36], Pinder and Grey [27], Gallagher et al. [17]). Unlike this last, the motion domain is divided into a large number of separate elements - the control volumes - and the weight functions acquire a unit value on one subdomain at a time and a null value outside. Since there is only one control volume surrounding each nodal point, the minimisation of the residues can be performed with reference to one subdomain Ω_I at a time (see fig.1).

Starting from the differential equation $\delta(\Xi) = 0$, the integrals involved in the process of minimisation of the residues are evaluated by expressing the variation of Ξ inside each subdomain by means of piecewise linear profiles between the grid points. The resulting discretisation equation expresses the conservation principle for Ξ in a control volume of finite dimensions, just as the starting differential equation expresses the conservation principle for one infinitesimal volume element. As a result, we get integral mass, momentum and energy conservation both on the individual control volume and on any group of control volumes and therefore also on the overall domain. Moreover, the exact integral conservation of these quantities does not depend on the size of the grid: it also holds good for elementary discretisations provided that care is taken with the procedures and the time step adopted for calculating the exchanges at the interfaces of the control volumes, since the time integration remains the finite difference type.

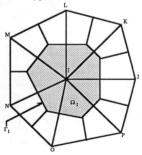

Fig. 1. Control Volume Finite Element domain discretisation.

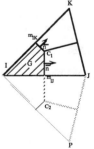

Fig. 2. Mesh elements involved in calculation of flow between node I and J.

By contrast, linear finite element schemes are unable to locally preserve mass and energy in problems in which flow exchange takes place (parabolics and hyperbolics in type) (De Marsily, [10]; Patankar, [26]).

As with the other numerical solution methods, a number of basic rules must be observed to ensure that the overall balance is satisfied and that the "physical" representation of the problem is maintained. As has been noted above, one of the main advantages of the control volume methods is the simplicity of interpreting the meaning of the terms appearing in the discretisation equation and, therefore, the possibility of checking immediately that the basic rules are satisfied. For example, the incoming and outgoing flows crossing the interface of two adjoining control volumes must have the same value. This property is reflected in the terms of the discretisation equation and determines the symmetry of the system matrices. As a last example, if the differential equation contains, at least in steady state conditions, just spatial derivatives of the dependent variable Ξ or $\Xi + \Xi_0$, in which Ξ_0 is a constant, they must still satisfy the differential equation. Transposition in the discretisation equation, which will be characterised by a system matrix whose diagonal terms will generally be the sum, in absolute value, of the extra-diagonal terms, is immediate. This latter property of the system matrix changes slightly in the case of unsteady state problems, depending on the particular method used to calculate the storage terms. This aspect will be examined in more detail below.

Domain discretisation. Once the position of the nodes in the domain has been identified, one may proceed to construct the triangular finite element mesh, of the linear type for the sake of simplicity, which, as is known, offers great flexibility in the reproduction of the domain and any discontinuities it may have. The second step entails identifying the boundaries of the control volumes bounded by the axes of the sides of the triangles of the mesh. In this manner a Dirichlet tessellation is obtained. It follows that the points C_1 e C_2 in fig. 2 correspond to the circumcentres of the two triangular elements. The exchanges between the nodes I and J are then calculated along the sides $\overline{C_1 \, m_{IJ}}$ (belonging to the element IJK) and $\overline{m_{IJ} \, C_2}$ (belonging to the element IPJ). The procedure followed in the construction of the polygons ensures maximum precision in calculating the exchanges since the boundaries of the control volume lie at the midpoint of the link between nodes I and J.

One final consideration: it may be noted that the structure of the system matrix (symmetrical, dominant diagonal, Stiltjes type) is similar to the one derived from the integrated finite difference approach, as can immediately be seen from an analysis of the similarities in the respective derivation processes, while in the case of the finite elements the matrix is no longer the Stiltjes type.

In order to illustrate the method in greater detail, the derivation of the discretisation equations in the case of overland flow is performed in the following section.

3. OVERLAND AND CHANNEL PROCESSES

The study of two-dimensional flow refers either to verified or possible flood event analysis or to the study of overland flow resulting from more or less intense rainfall. The mechanism governing overland and channel processes is more complicated than the one governing flow in pipelines due to the presence of a free surface the position of which may vary in time and space. In addition, there is an interdependence between water level, discharge, channel bed slope and free surface slope (Chow, [8], Dooge, [13]). The responses of the channel network vary considerably in time and space depending on the time variation of the flows entering the area in question and on the hydrogeometrical characteristics of the area.

The equations describing overland flow are the well-known De Saint Venant equations (Abbott, [1]):

$$\frac{\partial h}{\partial t} + \frac{\partial (uh)}{\partial x} + \frac{\partial (vh)}{\partial y} = q \tag{1}$$

$$\frac{\partial (uh)}{\partial t} + \frac{\partial \left(u^2 h\right)}{\partial x} + \frac{\partial (uvh)}{\partial y} + gh\left(S_{fx} + \frac{\partial H}{\partial x}\right) = 0 \tag{2}$$

$$\frac{\partial (vh)}{\partial t} + \frac{\partial \left(v^2 h\right)}{\partial y} + \frac{\partial (uvh)}{\partial x} + gh\left(S_{fy} + \frac{\partial H}{\partial y}\right) = 0 \tag{3}$$

Where:

$H(x,y,t)$ = water surface elevation;

$h(x,y,t)$ = local water depth;

t = time;

x, y = horizontal cartesian coordinates;

$u(x,y), v(x,y)$ = flow velocities in x and y directions (mean values in z direction);

$q(x,y,t)$ = net precipitation;

$S_{fx}(x,y), S_{fy}(x,y)$ = friction slopes in x and y directions;

g = gravity acceleration.

In accordance with Hromadka et al. [19], and Xanthopoulos et al. [34], the use of diffusive approximation - in which the inertial terms are ignored over the gravitational terms, frictions and pressure heads - permits the salient aspects of the phenomenon under study to be reproduced (cf. Hromadka, [18]; De Vries et al. [11], Labadie, [22]). Using this approach, the equations (2) and (3) can be replaced by the following system of parabolic differential equations:

$$\left(S_{fx} + \frac{\partial H}{\partial x}\right) = 0 \tag{4}$$

$$\left(S_{fy} + \frac{\partial H}{\partial y}\right) = 0 \tag{5}$$

In addition, applying the Manning/Stricker law to the description of the friction slopes that appear in the preceding equations, the relation between the velocity and water depth components can be obtained:

$$S_{fx} = \frac{n_x^2}{h^{4/3}} |\vec{w}| \, \vec{w} \times \vec{i} = \frac{n_x^2}{h^{4/3}} \sqrt{u^2 + v^2} \; u$$

$$S_{fy} = \frac{n_y^2}{h^{4/3}} |\vec{w}| \, \vec{w} \times \vec{j} = \frac{n_y^2}{h^{4/3}} \sqrt{u^2 + v^2} \; v$$

where $\vec{w} = u\vec{i} + v\vec{j}$, is the velocity vector and n_x, n_y, are the Manning roughness coefficients in directions x and y respectively and, therefore:

$$|\vec{w}|^2 = u^2 + v^2 = \sqrt{\left(\frac{S_{fx}^2}{n_x^2} + \frac{S_{fy}^2}{n_y^2} \right)} \, h^{4/3}$$

Lastly, the replacement of the preceding relation in the expression of S_{fx} and S_{fy}, determines the following expressions for the components of the velocity vector:

$$u = -\sqrt{\frac{\partial H}{\partial x}} \, \frac{h^{2/3}}{n_x^2} \, \frac{1}{\sqrt[4]{\left(\frac{\partial H}{\partial x} \right)^2 \frac{1}{n_x^4} + \left(\frac{\partial H}{\partial y} \right)^2 \frac{1}{n_y^4}}} \tag{6}$$

$$v = -\sqrt{\frac{\partial H}{\partial y}} \, \frac{h^{2/3}}{n_y^2} \, \frac{1}{\sqrt[4]{\left(\frac{\partial H}{\partial x} \right)^2 \frac{1}{n_x^4} + \left(\frac{\partial H}{\partial y} \right)^2 \frac{1}{n_y^4}}} \tag{7}$$

Equations (6) and (7) differ from the ones normally used (cf. Hromadka et al., [18]; Reitano, [28] and [29]) in that they have been derived on the assumption of one-dimensional flow in anisotropic conditions and considering the x and y axes to coincide with the main anisotropy directions of the propagation domain under study.

4. DISCRETISED EQUATIONS IN OVERLAND FLOW

The replacement of equations (6) and (7) in equation (1), on the assumption that x and y are the main anisotropy directions, leads to the following system of partial derivative differential equations of the parabolic type:

$$\frac{\partial}{\partial x}\left[k_x \frac{\partial H}{\partial x} \right] + \frac{\partial}{\partial y}\left[k_y \frac{\partial H}{\partial y} \right] + q = \frac{\partial H}{\partial t} \tag{8}$$

$$k_x = \frac{1}{n_x^2} \frac{1}{\varphi(\nabla H)} h^{5/3} \qquad k_y = \frac{1}{n_y^2} \frac{1}{\varphi(\nabla H)} h^{5/3}$$

$$\varphi(\nabla H) = \left[\left(\frac{\partial H}{\partial x} \right)^2 \frac{1}{n_x^4} + \left(\frac{\partial H}{\partial y} \right)^2 \frac{1}{n_y^4} \right]^{1/4}$$

(9)

Once this initial and boundary conditions have been assigned (prescribed head, prescribed discharge or mixed conditions), the problem is fully defined in mathematical terms and the solution is obtained by the integration of equation (8).

The relevant discretisation equation will be obtained with reference to just the node I and its control volume (Ω_I); extension to the overall domain (Ω) will be accomplished by assembling the contributions for all the nodes. The procedure illustrated is legitimate insofar as the control volumes constitute a partition of the whole domain.

The first step in the application of the method is the integration of equation (8) on the surface domain Ω_I bounded by the curve Γ_I (see fig. 1 for the meaning of the symbols):

$$\int_{\Omega_I} \left(\frac{\partial}{\partial x} \left[k_x \frac{\partial H}{\partial x} \right] + \frac{\partial}{\partial y} \left[k_y \frac{\partial H}{\partial y} \right] \right) d\Omega_I + \int_{\Omega_I} q \, d\Omega_I = \frac{\partial}{\partial t} \int_{\Omega_I} H \, d\Omega, \quad (10)$$

The application of the divergence theorem and the reformulation of the term on the right of the equals sign gives:

$$\underbrace{\int_{\Gamma_I} \left[\left(k_x \frac{\partial H}{\partial x} \right) \vec{i} + \left(k_y \frac{\partial H}{\partial y} \right) \vec{j} \right] \times \vec{n}_{\Gamma_I} d\Gamma_I}_{\Im} + q_I \Omega_I = \frac{\partial V_I}{\partial t} \quad (11)$$

where $\times$ signifies a scalar product; $\vec{i}$ and $\vec{j}$ are unit vectors of the x and y coordinates; $\vec{n}_{\Gamma_I}$ is the unit normal to the line Γ_I, V_I is the volume of water contained in the control volume to which node I belongs, $\partial V_I / \partial t$ is the volume time variation and q_I is the mean value of the outside discharge entering the surface domain Ω_I.

In equation (11) the line integral $\Im$ represents the total discharge crossing the boundary Γ_I of the domain Ω_I for all the elements sharing the same node I, while the second term, to the right of the equals sign, represents the exchanges with the outside. Since the triangles surrounding node I contain portions of the control volume and its boundary, the total value of the exchange can be obtained on the basis of the contribution furnished by the portion G of the control volume indicated in figure 2. Linear functions of the following type are used as shape functions:

$$\Phi_i(x,y) = \frac{1}{2A}\left(a_i y + b_i x + c_i\right) \qquad i = I, J, K$$

$$a_k = x_j - x_i$$
$$b_i = y_j - y_k \qquad\qquad i, j, k = I, J, K$$
$$c_i = x_j y_k - x_k y_j$$

where A is the area of the triangle IJK and a_k and b_i are obtained from anticyclic permutations of the x coordinates and cyclic permutations of the y coordinates respectively. Each variable defined inside the triangular element may in this way be assigned on the basis of its known value at the nodal points. For example, the head inside the triangular element may be expressed by the relation:

$$H(x,y) = \sum_i \Phi_i(x,y)\, H_i \qquad i = I, J, K \qquad (12)$$

In the event that $n_x = n_y = n$ and, therefore, $k_x = k_y$, the analytical solution of the line integral in (11), limited to just the element IJK, gives the value of the discharge crossing the boundaries $\overline{C_1\, m_{IJ}}$ and $\overline{C_1\, m_{IK}}$ (see fig. 2):

$$Q_I = -\left[(T_{IJ} + T_{IK})H_I - T_{IJ}H_J - T_{IK}H_K\right]$$

$$T_{IJ} = \gamma(I,J)\frac{1}{\varphi(\nabla H)}\frac{1}{n}\frac{h_{c_1}^{8/3} - h_{m_{IJ}}^{8/3}}{h_{c_1} - h_{m_{IJ}}}$$

$$\left.\begin{array}{l} T_{IJ} \geq 0 \\ T_{IK} \geq 0 \end{array}\right\} \forall I, J, K \qquad (13)$$

$$T_{IK} = \gamma(I,K)\frac{1}{\varphi(\nabla H)}\frac{1}{n}\frac{h_{c_1}^{8/3} - h_{m_{IK}}^{8/3}}{h_{c_1} - h_{m_{IK}}}$$

where $\gamma(I,J)$ and $\gamma(I,K)$ are functions of the positions of the nodes I, J and K. If the starting triangle is acute, the preceding functions always have a positive value, thus ensuring the positive value of the coefficients of the transmissivity matrix ($\mathbf{T}$), which is required by the method and is also necessary, as has been noted, in order for the physical significance of the problem to be respected. This latter result is valid in isotropic situations, while in anisotropic conditions slightly different results apply.

If we also consider the flows for nodes J and K, the expression containing the contributions for one single element in the mesh can be obtained in matrix form:

$$
\begin{vmatrix} Q_I \\ Q_J \\ Q_K \end{vmatrix} = - \begin{vmatrix} (T_{IJ}+T_{IK}) & -T_{IJ} & -T_{IK} \\ -T_{IJ} & (T_{IJ}+T_{JK}) & -T_{JK} \\ -T_{IK} & -T_{JK} & (T_{IK}+T_{JK}) \end{vmatrix} \begin{vmatrix} H_I \\ H_J \\ H_K \end{vmatrix} \qquad (14)
$$

Note that the transmissivity matrix is symmetrical and that it therefore satisfies the condition laid down by the physical problem: equality of the discharge flowing between node I and node K and between node K and node I. In addition, the diagonal terms have the property of assuming a value equal to the sum, in absolute value, of the extra-diagonal terms. With regard to this aspect, the discretisation equations obtained with the CVFEM are identical to those deriving from the IFDM except for the transmissivity coefficient calculation method which, in this case, is performed on the basis of the properties of the triangular elements.

The last step entails the assembly of all the elements connected to node I and, lastly, the assembly of all the nodes in the domain, obtaining the following expression for the exchanges between all the nodes in the domain:

$$
\{Q\} = \begin{vmatrix} Q_1 \\ Q_2 \\ \vdots \\ Q_{n-1} \\ Q_n \end{vmatrix} = - \begin{vmatrix} \displaystyle\sum_{i=2}^{n} T_{1i}^g & -T_{12}^g & \cdots & -T_{1(n-1)}^g & -T_{1n}^g \\ -T_{12}^g & \displaystyle\sum_{\substack{i=1 \\ i\neq 2}}^{n} T_{2i}^g & \cdots & -T_{2(n-1)}^g & -T_{2n}^g \\ \cdots & \cdots & \cdots & \cdots & \cdots \\ -T_{1(n-1)}^g & -T_{2(n-1)}^g & \cdots & \displaystyle\sum_{\substack{i=1 \\ i\neq n-1}}^{n} T_{(n-1)i}^g & -T_{(n-1)n}^g \\ -T_{1n}^g & -T_{2n}^g & \cdots & -T_{(n-1)n}^g & \displaystyle\sum_{\substack{i=1 \\ i\neq n}}^{n} T_{ni}^g \end{vmatrix} \begin{vmatrix} H_1 \\ H_2 \\ \vdots \\ H_{n-1} \\ H_n \end{vmatrix}
$$

$$
= -[\mathbf{T}]\{\mathbf{H}\}
$$

where T_{ij}^g denotes the value taken by T with reference to the generic nodes I and J and to the two elements that share the IJ side.

As far as the calculation of the unsteady state term is concerned, two calculation methods, known as the lumped and consistent formulation, can be used. Generally speaking, the volume contained inside the domain to which the node I belongs is a function of the head at all the nodes surrounding node I:

$$V_I = f(H_I, H_\ell) \quad \ell = \{\text{set of nodes surrounding node I}\}$$

and, therefore:

$$\frac{\partial V_I}{\partial t} = \frac{\partial V_I}{\partial H_I} \frac{\partial H_I}{\partial t} + \sum_\ell \frac{\partial V_\ell}{\partial H_\ell} \frac{\partial H_\ell}{\partial t}$$

In the consistent formulation, whose approach is based on the preceding expression, the variation in volume at node I depends on the head variations at all the nodes surrounding node I and the relevant contribution to the system matrix is present in all its non-null elements. Unfortunately, although it provides for a more accurate calculation of the storage term, the consistent formulation results in the loss of dominance of the system matrix diagonal. In the lumped formulation the second member summation is usually ignored and the term that takes account of the storage at the node is concentrated solely in the diagonal element of the matrix. The system matrix is therefore of the dominant diagonal type (the extra-diagonal terms are in fact all negative). This system matrix structure offers major advantages in computational terms since it permits the use of rapid convergence solution algorithms for the algebraic equations system. A similar computational advantage of the lumped formulation over the consistent type lies in the integration methods based on the traditional finite element approach. Neuman [24], in the analysis of a problem regarding flow in an unsaturated porous medium conducted with finite element integration based on the Galerkin method, encountered convergence problems in the case of consistent approximation: these problems were eliminated when the lumped approach was used. Huyakorn and Pinder [20] also report that the lumped approach, in flow in porous media, generally produces less accurate but more stable solutions. Recently, Wood and Calver [35] compared the results derived from the two approaches and did not find any substantial differences in convergence, though they noted a considerable increase in the accuracy of the solution obtained using the consistent approach.

In order to enhance the reproduction of the storage term, while retaining the advantages of the lumped formulation, a correction was developed in the method for calculating the variation in the volume stored in the element. The basic idea is to evaluate correctly the storage terms at the nodes (i.e. according to the levels at all the connected nodes) but to assign these values solely to the diagonal term of the system matrix. For the purposes of the calculation, the variation over time of the volume contained in the domain to which node I belongs is expressed in the form:

$$\frac{\partial V_I}{\partial t} = \Omega'_I (H\ell)\frac{\partial H_I}{\partial t} \qquad \ell = \{\text{set of nodes surrounding node I }\}$$

where the control volume surface domain is no longer regarded as fixed but as varying with changes in the piezometric head of the nodes surrounding the node under study.

The stored volume calculation procedure used here proves particularly useful in the case of depletion of the elements and, though it generates a limited increase in the number of iterations and therefore increases computation time, it achieves greater precision in the evaluation of the levels and the exchanges between the nodes. In fact, particularly in the depletion phase, it can happen that the domains are not uniformly filled thus not respecting the basic hypothesis of the lumped approach.

The time discretisation was originally performed using a centred implicit scheme (Crank-Nicholson with $\alpha = 0.5$). The implicit scheme was preferred over the explicit type since, as is known, it permits the use of larger time discretisation steps without causing problems of stability. Accordingly, we obtain:

$$\left[\alpha[\mathbf{T}] + \left[\frac{\Omega'}{\Delta t}\right]\right]\{\mathbf{H}\}^{t+\Delta t} = \left[-(1-\alpha)[\mathbf{T}] + \left[\frac{\Omega'}{\Delta t}\right]\right]\{\mathbf{H}\}^t +$$

$$+ [\Omega]\left((1-\alpha)\{q\}^t + \alpha\{q\}^{t+\Delta t}\right) \qquad (15)$$

where Δt is the time step used in the simulation.

In order to step up the convergence towards the system solution it was decided to furnish an estimation, at step start, of the value of the water levels. Different methods for predicting the head value, based on interpolation and extrapolation techniques, were analysed. The comparison was made with the obtainable results without the help of any forecasting mechanism regarding either the correctness of the value obtainable at step end or the number of iterations required to achieve the solution. The results of the tests showed that the greatest reduction in the number of iterations is obtained by using an estimation of the value at step start based on a parabolic extrapolation method which takes account, at the n-th time step, of the solutions obtained at the (n-3)-th, (n-2)-th and (n-1)-th steps. In the case of linear forecasting, the average number of iterations is reduced by approximately 20% while, where parabolic forecasting is adopted, a reduction of approximately 40% in the number of iterations is achieved (with peaks of approximately 60%).

The advantages deriving from the introduction of a quadratic method in the approximation of time derivatives compared with the traditional Crank-Nicholson linear model are currently being evaluated. In theoretical terms, the advantages of this method lie mainly in a correct approximation up to the second order compared with a linear approximation, such as the Crank-Nicholson one, even though it has second order terms that become negligible.

The assigning of the starting and boundary conditions is reflected in a partition of the heads vector at the nodes: some of the values of H are in fact assigned and therefore combine to form part of the known term of equation (15). This equation is then transformed into a system of equations which, expressed in matrix form, becomes:

$$[B]\{H\}^{t+\Delta t} = [A]\{H\}^{t} + \{C\} \tag{16}$$

The problem of the integration of equation (8) has therefore been transformed into the search for the solution to the non-linear system of algebraic equations in the unknowns H. If n_t is the total number of nodal points in the domain and n_0 is the number of nodes in which the head is assigned, the vector H will have $(n_t - n_0)$ unknowns and the system matrix will be a sparse matrix measuring $(n_t - n_0) \times (n_t - n_0)$.

The non-linearity of the starting equation calls for the recursive calculation of the values of T (functions of the local value of h) and suggests the adoption of iterative solution methods. At each time step a certain number of equation systems must therefore be resolved and the efficiency of the solvers to be used for the system (16) becomes crucial. The knowledge of the topological structure of the system matrix must therefore be used to the fullest advantage: (1) The solver must take into account sparsity, symmetry and positive definiteness; (2) The method can be iterative; (3) The modified conjugate gradient method is found to be the most efficient with respect to memory requirements (Ajiz et al., [5]; Todini et al., [31]); (4) Some direct methods are the most efficient from the point of view of computational time requirements.

At the present time it has been found that the most suitable solution methods are the modified conjugate gradient method and a direct method developed by I.S. Duff and J.K. Reid - available in Harwell bookshops (MA27) -which is particularly effective for solving systems of algebraic equations with sparse system matrix and large dimensions. With regard to this latter method, it may be noted that the direct methods based on a resorting of the system matrix save computational time and, though they require a larger memory allocation than the iterative methods, they require only a limited increase in memory storage and are therefore suitable for solving problems of major proportions (Duff,[14] and Duff et al. [15])

5. NUMERICAL RESULTS

In this initial phase, which is entirely given over to a verification of the computation programme developed, the mathematical model was tested to compare its results with those obtainable with other numerical solution methods and with those deriving from an analytical solution. The possibility of using an analytical solution limited the choice of possible cases. The case chosen and illustrated below is not ideally suited to highlighting the advantages of the method, but the discussion furnished in the preceding chapters has already pointed up these benefits and briefly delineated its vast range of potential application. The simulated case, whose domain is illustrated in figs. 3 and 4, refers to a V-shaped catchment in which propagation was studied both in the case of a catchment characterised by a single slope in the direction of the channel and in the case of a double slope. As has been said, solutions obtained with other integration methods are available for this catchment and an approximated analytical solution can also be found in the literature (cf. Overton and Brakensiek, [25] and Stephenson and Meadows [30] for further details)

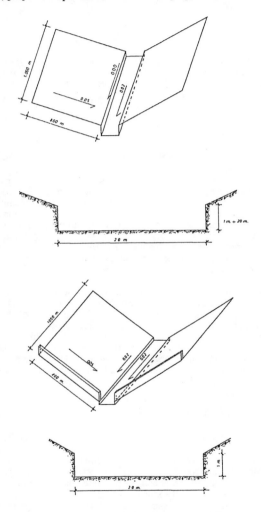

Fig. 3 and 4. Level V-catchment and tilted V-catchment.

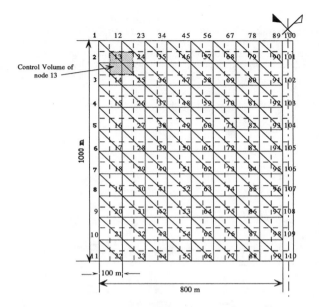

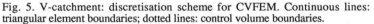

Fig. 5. V-catchment: discretisation scheme for CVFEM. Continuous lines: triangular element boundaries; dotted lines: control volume boundaries.

Since in the single slope V-catchment there is no gradient in the valley direction, the discharge flows perpendicularly to the channel. By contrast, in the dual-slope V-catchment both the valley side and the channel are characterised by a slope extending from upstream to downstream: in this case overland flow is diagonal to the channel. Propagation in the channel was managed together with propagation in the plain, though it was assumed to be one-dimensional and the channel propagation equations were discretised using an integrated finite difference approach. This dual equation discretisation method was made possible thanks to the above-mentioned similarities in the system matrices deriving from the control volume finite element method and the integrated finite difference method. The simultaneous management of different methods is another advantage of the scheme used in that it simplifies the solving of propagation problems that are markedly one-dimensional and which one does not wish to describe by two-dimensional approaches.

In the control volume finite element approach to both the single slope V-catchment and the twin slope catchment, a discretisation with "uniform" grid was used together with control volumes measuring 100x100 m in most cases. The mesh in the channel was given a rectangular shape (20 x 100 m). Both of the spatial discretisation methods are illustrated in figure 5. The simulation was conducted further to a rainfall input with an intensity of 10.8 mm/h and a duration of 1.5 h. The precipitation was assumed not to affect the channel. The attribution of the parameters to the catchment, though unrealistic in natural conditions, was undertaken so as to allow a comparison with results available from the analytical solution and with a solution using the SHE model based on a conventional finite difference scheme (Bathurst, [6]). The values of the parameters used are given in Table I. Similar considerations apply to the values chosen for the dimensions and slopes of the valley and channel slopes.

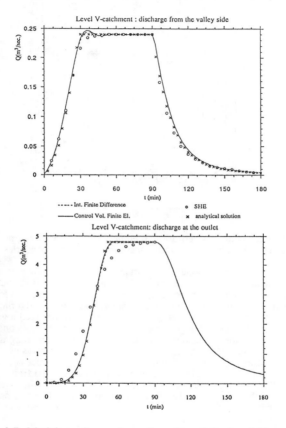

Fig. 6 and 7. Model results on the valley side of the level V-catchment; Discharge at the outlet in the level V-catchment.

Tab I

$n_v = 0.015$	Manning's resistance coefficient for the valley side
$n_c = 0.15$	Manning's resistance coefficient for the channel

Figs. 6 and 7 show a comparison between the kinematic analytical solution, the numerical solution obtained with the SHE model, an integrated finite difference model and this model in the single slope catchment configuration. The results for the SHE model were obtained using a grid of squares with a side length of 100 m. To calculate flow in the plain, the scheme followed was explicit, while an implicit scheme was used for the simulation of propagation in the channel. (Abbott et al., [2],[3] and [4]; Clausen, [9]). The discretisation scheme used in the IFDM employs the same type of mesh.

Note that there is no comparison with the results derived from the kinematic analytical solution at the catchment outlet section for the descending wave phase. This is due to the fact that the analytical solution adopted requires an instantaneous interruption in the channel inflow, while in this case the inflow from the plain diminishes gradually.

As can be seen in figures 6 and 7, the results obtainable with the CVFE method and the IFD scheme are very close to those of the kinematic analytical solution. An improvement can also be noted over the solution obtained with the finite difference scheme using the SHE model, as reported in Bathurst [6]. These observations apply in particular to the channel's outflow in the case of the single slope catchment. With regard to the double slope catchment, there is an increase in water level at the impermeable boundary of the catchment. This "barrier" is responsible for the differences between the solutions obtained with the IFD and CVFE schemes. In fact, in the latter scheme the nodes are positioned exactly at the boundary and are therefore more sensitive to a rise in the water level. In the IFD scheme the different position of the nodes, at the centroids of the grid elements, obscures this effect. The results obtained are shown in figure 8 and are compared with those obtained with the SHE model.

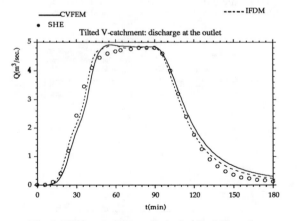

Fig. 8. Discharge at the outlet in the tilted V-catchment.

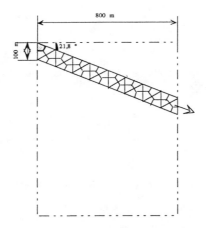

Fig. 9. Discretisation scheme for the comparison of CVFEM, IFDM and analytical results in the tilted V-catchment.

A more correct comparison with the approximated analytical solution in the double slope catchment can be obtained by the simulation for a single strip in the maximum slope direction of the catchment (see fig. 9). The CVFEM, IFDM and the analytical solution for the discharge flowing from the plain into the channel show in this case that the control-volume finite element scheme achieves greater accuracy in the reproduction of the two-dimensional nature of the flow compared with the integrated finite difference solution. It can in fact be seen in figure 10 that the CVFEM solution is very close to the analytical solution.

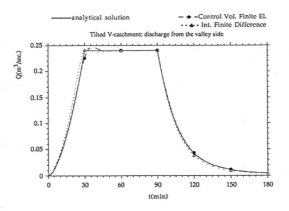

Fig. 10. Results for a strip of the tilted valley side.

6. CONCLUSIONS AND FUTURE LINES OF RESEARCH

The method presented here overcomes a number of integration problems that may arise when using different schemes in the integration of two-dimensional propagation equations.

Following the verification of the calculation program, accomplished by comparison with a theoretical solution found in literature and with results deriving from different numerical schemes, the method will be applied to the solution of 2-D overland flow problems instead of the IFDM previously used (see Di Giammarco et al., [12]).

The proposed method, not only maintains the simplicity of the IFDM even in dealing with plain-channel exchanges (two-dimensional - one-dimensional), it also allows a better representation of the complex variability of gradients and of friction coefficients which characterise real problems.

This study was developed within the framework of the research and development contract between the Commission of the European Communities and the University of Bologna, project no. EPOCH-CT90-0023.

REFERENCES

1. Abbott, M.B. Computational hydraulics - Elements of the theory of free surface flows. Pitman, London, 1979.

2. Abbott, M.B., Andresen, J.K., Havnø, K., Kroszynski, U.I. And Warren, I.R. Research and development for the unsaturated zone component of the European Hydrologic System - Système Hydrologique Européen (SHE), In M.B. Abbott and J. Cunge (Editors), Engineering Applications of computational Hydraulics, Pitman, London, 1: 40-70, 1982.

3. Abbott, M.B., Bathurst, J.C., Cunge, J.A., O'Connell, P.E. and Rasmussen, J. An Introduction to the European Hydrological System - Système Hydrologique Européen "SHE", 1 History and philosophy of a physically based, distributed modelling system. J. Hydrol., 87: 45-59, 1986.

4. Abbott, M.B., Bathurst, J.C., Cunge, J.A., O'Connell, P.E. and Rasmussen, J. An Introduction to the European Hydrological System - Système Hydrologique Européen "SHE", 2 Structure of a physically based, distributed modelling system. J. Hydrol., 87: 61-72, 1986.

5. Ajiz, M., Jennings, A. A robust incomplete Cholesky conjugate gradient algorithm, Int. Jour. for Numerical Methods in Engineering, Vol. 20, pp. 949-966, 1984.

6. Bathurst, J.C. Preliminary tests of the SHE with a V-catchment, SHE Report No. 23, 1983.

7. Bechteler, W., Kulish, H., Nujic, M. 2-D dam-break flooding waves comparison between experimental and calculated results, in Flood and flood management ed. by J.Saul, Kluwer Academic Publishers, Dodrecht, 1992.

8. Chow, V. Handbook of applied hydrology. McGraw Hill, New York, 1964.

9. Clausen, T.J. SHE - A Short Description. Danish Hydraulic Institute, Hørsholm, 1979.

10. De Marsily G. Quantitative hydrogeology - Groundwater hydrology for engineers. Academic Press, Inc., Orlando, 1986.

11. De Vries, J.D., Hromadka, T.V., Nestlinger, A.J. Applications of a two-dimensional Diffusion Hydrodynamic model, Hydrosoft 86 - Hydraulic Engineering Software, Proc. 2nd Int. Conf. Southampton, 393-412, 1986.

12 Di Giammarco, P., Todini, E., Consuegra, D., Joerin, F., Vitalini, F. Combining a 2-D flood plain model with GIS in order to assess flood damages., same book.

13. Dooge, J.C.I. Theory of flood routing. in: River flow modelling and forecasting ed by Kraijenhoff D.A. and Moll, J.R., D, Reidel Publishing Company, Dordrecht, 1986.

14. Duff, I. The solution of nearly symmetric sparse linear equations. Sixth International Conference on Computing Methods in Applied Sciences and Engineering, Versailles, 1983.

15. Duff, I., Erisman, A.M., Reid, J.K. Direct methods for sparse matrices. Oxford University Press, London, 1986,

16. Fung, L.S.K., Hiebert, A.D., Nghiem, L. Reservoir simulation with a control-volume finite-element method. 11th SPE Symposium on Reservoir Simulation, Anaheim, California, February 17-20, 231-242, 1991.

17. Gallagher, R.H., Oden, J. T., Taylor, C., Zienkiewicz, O.C. Finite elements in fluids, vol. 1, Viscous flow and hydrodynamics, J. Wiley & Sons, London, 1975.

18. Hromadka, T.V., Berenbrock, C.E., Freckleton, J.R., Guymon, G.L. A two-dimensional dam-break flood plain model. Adv. Water Resources, vol. 8, March, 7-14, 1985.

19. Hromadka II, T.V., and Guymon, G.L. Nodal domain integration model of two-dimensional diffusion processes, Adv. Water Resources, 1981.

20 Huyakorn, P.S. and Pinder, G.F. Computational methods in subsurface flow, Academic Press, Orlando, 1983.

21. Katopodes, N.D. Two-dimensional surges and shocks in open channels., J. Hydr. Div. ASCE, 110(HY6), 794-812, 1984.

22. Labadie,G. Flood waves and flooding models. Pre-proceedings of the NATO-ASI on "Coping with floods", Erice, November 3-15, 1992.

23. Narasimhan, T.N., Witherspoon, P.A. An integrated finite differences method for analyzing fluid flow in porous media. Water Res. Research, 12 (1), 57-64, 1976

24. Neuman, S.P. Galerkin approach to saturated-unsaturated flow in porous media, in Finite Elements in fluids, vol.1, ed by R.H. Gallagher, ch. 10, John Wiley, New York, 1975.

25. Overton, D.E., and Brakensiek, D.L. A kinematic model of surface runoff response. Proc. IAHS-UNESCO Wellington (NZ) Symp. on the Results of Research on Representative and Experimental Basins. IAHS Publ. No. 96, 110-112, 1973.

26. Patankar, S.V. Numerical heat transfer and fluid flow. Hemisphere Publishing Corporation, New York, 1980.

27. Pinder, G. F., Gray, W.G. Finite element simulation in surface and subsurface hydrology., Academic Press, London, 1977.

28. Reitano B., Modello bidimensionale per la simulazione di inondazioni fluviali, in Atti del 23° convegno di Idraulica e Costruzioni Idrauliche, Firenze, 31 Agosto-4Settembre, D.321-D.336 (In Italian), 1992.

29. Reitano B., Flooding vulnerability analysis at basin-wide scale. Pre-proceedings of the NATO-ASI on "Coping with floods", Erice, November 3-15, 1992.

30. Stephenson, D., Meadows, M.E. Kinematics hydrology and modelling. Elsevier, Amsterdam, 1986.

31. Todini, E. and Pilati, S. A gradient algorithm for the analysis of pipe networks., Proc. Int. Conf. on Computer Applications for Water Supply and Distribution, Volume 1 (System analysis and simulation), John Wiley & Sons, London, 1987.

32. Todini, E. and Venutelli M. Overland flow: a two-dimensional modelling approach., In D.S. Bowles and P.E. O'Connell (Ed.), Recent Advanced in the Modeling of hydrologic systems, Kluwer, Dodrecht, 153-165, 1988.

33. Todini, E., Venutelli M. Overland flow: a two-dimensional modelling approach., Proc. NATO Advanced Studies Institute, Workshop on recent advances in Hydrology, 1990.

34. Xanthopoulos, Th. and Koutitas, Ch. Numerical simulation of a two-dimensional flood wave propagation due to dam-failure, J. of Hyd. Res., vol. 14, no 4, 321-331, 1976.

35. Wood, W:L:, Calver, A. Lumped versus distributed mass matrices in the finite element solution of subsurface flow., Water Resources Res., vol. 26, no. 5, 819-825, 1990

36. Zienkiewicz, O.C., The finite element method in Engineering science. McGraw Hill, London, 1971.

Finite Element Algorithms for Modelling Flood Propagation

J.-M. HERVOUET*, J.-M. JANIN**
* Fluvial Hydraulics Group
** Maritime Hydraulics Group
Laboratoire National d'Hydraulique
Direction des Etudes et Recherches, Electricité de France
6 Quai Watier 78401 Chatou Cedex

INTRODUCTION:

Flood propagation over initially dry areas offers a challenging topic for numerical simulation. Many physical phenomena such as hydraulic jumps, transcritical flows and turbulence, result in numerical problems. The treatment of dry areas is another difficulty. Finally, a complex geometry including minor bed and flood plain over long reaches must be handled. One dimensional Finite Differences codes have long been the standard analytical tool for such studies but 2D Finite Element programmes now offer a viable alternative. The assets of the theoretical background of Finite Elements and their flexibility for complex domains are generally accepted but the difficulty of implementation and the computational time are still often criticised. This argument is now vanishing since considerable progress has been made in recent years with Element By Element techniques [4], new upwinding methods and new computer architecture. At L.N.H. Finite Element codes are now faster than Finite Difference codes on curvilinear grids and the latter are being progressively replaced. In this paper we shall detail the numerical solutions adopted in the TELEMAC-2D code, that solves the shallow water equations (ref. [1],[2],[3]). The numerical choices have been driven by the necessity to cope with the physical phenomena already mentioned and the need for a computational efficiency to treat very large domains. Among the applications are flood propagation, dam break simulations and flow in estuaries. In these applications propagation over initially dry areas and the drying ot wet areas raise the most difficult problems.

CHOICE OF ALGORITHMS:

First of all, let us say that by restricting this paper to Finite Elements we do not infer that other techniques such as Finite Differences or Finite Volumes are of a lesser interest. Actually those techniques offer extra possibilities for designing numerical schemes, especially for advection equations. For example efforts for adapting Finite Volumes schemes to non-structured meshes are a very interesting approach. However, as far as we are concerned and given the industrial use of our programmes, we think that non structured meshes and unconditionally stable schemes (i.e. implicit or semi-implicit) are a major asset. If we take for granted the choice of Finite Elements, the main remaining alternatives are the formulation of the equations (conservative, non-conservative,..), the choice of the numerical schemes, especially for advection terms, and the treatment of wetting and drying areas (moving boundary techniques).

The choice of the formulation of shallow water equations is not obvious and a conservative form (i.e. discharge-depth) seems a priori better, however in such a formulation divisions by the depth are needed to get the velocity field and this generally leads to nasty problems in drying zones. This formulation has been implemented in Finite Differences in a former programme at LNH and it has been abandoned at the time being. Moreover stability analysis of various numerical schemes [5] are in favour of non conservative equations. It must be noted that in Finite Elements mass-conservation can be ensured with non-conservative equations. Moreover, with shock capturing techniques (i.e. without special treatment of discontinuities), as there is no discontinuity in the finite element approximations, the two formulations are equivalent.

Non-conservative equations have thus been chosen for TELEMAC and two options have been developed : the celerity-velocity and the depth-velocity options. The equations of the first option read:

$$\frac{\partial c}{\partial t} + \vec{u}.\overrightarrow{grad}(c) + \frac{c}{2}\,div(\vec{u}) = 0$$

$$\frac{\partial u}{\partial t} + \vec{u}.\overrightarrow{grad}(u) + 2c\,\frac{\partial c}{\partial x} - div(\,v_t\,\overrightarrow{grad}(u)\,) = S_x - g\,\frac{\partial Z_f}{\partial x}$$

$$\frac{\partial v}{\partial t} + \vec{u}.\overrightarrow{grad}(v) + 2c\,\frac{\partial c}{\partial y} - div(\,v_t\,\overrightarrow{grad}(v)\,) = S_y - g\,\frac{\partial Z_f}{\partial y}$$

with the following notation :

u,v   :   velocity components.

c   :   celerity of shallow water waves.

z_f :   bottom level.

S_x, S_y :   source/sink terms (bottom friction, wind, etc.).

g   :   gravity acceleration.

v_t :   eddy viscosity.

The more classical depth-velocity formulation reads, h being the depth:

$$\frac{\partial h}{\partial t} + \vec{u} \cdot \overrightarrow{\text{grad}}(h) + h \, \text{div}(\vec{u}) = 0$$

$$\frac{\partial u}{\partial t} + \vec{u} \cdot \overrightarrow{\text{grad}}(u) + g \frac{\partial h}{\partial x} - \text{div}(v \, \overrightarrow{\text{grad}}(u)) = S_x - g \frac{\partial Z_f}{\partial x}$$

$$\frac{\partial v}{\partial t} + \vec{u} \cdot \overrightarrow{\text{grad}}(v) + g \frac{\partial h}{\partial y} - \text{div}(v \, \overrightarrow{\text{grad}}(v)) = S_y - g \frac{\partial Z_f}{\partial y}$$

The celerity-velocity option has elegant properties of symmetry and conditioning of the linear systems but the depth-velocity is more convenient to design mass-conservative schemes. However these two options are very similar and the final choice remains an open question.

Advection terms appear in all the equations, including the continuity equation, and the upwinding is very important, especially with high Froude numbers. Several schemes have been tested and currently the following solution has been adopted:

- S.U.P.G. [6] is used for the continuity equation
- The characteristics method is used for the momentum equation.

A fractional step approach is thus used for the momentum equation and, after advection, a second step deals with all the remaining terms.

A semi-implicit scheme, including S.U.P.G., has been chosen to achieve an unconditional stability. By semi-implicit scheme we mean that, except for the time derivative terms, unknowns h, u and v are discretised in the form $h = \theta h^{n+1} + (1-\theta)h^n$, etc., where θ is an implicitation coefficient. The stability obtained allows us, in some cases, to perform computations with Courant numbers up to 50. However, despite the unconditional stability, one must keep in mind that the Courant number is a physically based number that must be taken into account to get good quality solutions, especially in the cases where the advection is dominant. Moreover, iterative techniques are used for solving the linear system stemming from the finite element discretisation and the optimum time step for a given simulation (as regards computational time) is not necessarily the highest.

Many of the recent studies performed with TELEMAC involve wetting and drying areas and it seems a general trend linked to the increasing number of environmental problems resorting to hydraulics. The ability to cope with such situations is thus of utmost importance. Two kinds of solutions can be considered:

* solving the equations everywhere.
* removing the drying zones from the computational domain.

The first one is the simplest but corrections must be applied in the wetting and drying zones, to avoid spurious values of the free surface gradient. In dry areas, this gradient is equal to the gradient of the bottom topography and in that case must not act as a driving force in the momentum equation. This problem of spurious free surface gradient is exemplified in 1D on the sketch below:

Free surface in the sense of finite elements

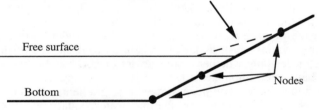

Free surface

Bottom

Nodes

The second option, often referred to as "moving boundary technique", consists of removing the drying zones from the domain. This can be a very cumbersome task and might require the definition of new meshes, leading to heavy operations to be performed at every time-step, e.g. the renumbering of the unknowns, and thus spoiling the efficiency. To avoid this in TELEMAC-2D, dry elements are kept in the mesh, but their occurrence in all the algorithm is cancelled by an array that is set to 0 for dry elements and 1 for the others. A thorough study of all these algorithms showed that one could restrict this operation to a very small number of routines, mainly those with an assemblage.

RESULTS:

In this conference, another paper called "Computation of a flood event using a two dimensional finite element model and its comparison to field data" deals with a validation of TELEMAC-2D in a 10 km reach of the river Culm in Devon. Very large domains may indeed now be treated in finite elements, either on super computers or workstations. Figures 1 and 2 show respectively a 28000 elements mesh and the inundation extent in a very complex area flooded after a dyke break. This computation has been performed on a workstation.

The two kinds of treatment for dry domains, removing elements or solving the equations everywhere, have been implemented in TELEMAC. Figures 3 and 4 show an example of flow in an estuary with tidal flats. Results are not significantly different but they are slightly better with the second technique ; however some problems arise: depending on the criterion for detecting dry elements mass-conservation is not ensured exactly, furthermore the propagation of a flood may also be limited by the process of reactivating elements.

Solving the equations everywhere remains at the time being the standard solution in TELEMAC-2D. With this kind of approach, the quality of the results depends greatly on the type of correction applied and on the algorithms used to detect the half dry/half wet elements. These algorithms must, as far as possible, avoid the use of threshold values for deciding whether a point is dry or not. For the two methods which have been implemented, some problems may happen when the drying zones have very steep slopes, e.g. gradients greater than 1 ; however high this value for shallow water equations, it must be considered that none of these methods is fully satisfactory at the moment and research is still going on, especially on the criteria for detecting dry elements.

PROPAGATION OVER DRY DOMAINS IN 3D:

3D computations of flows with wetting and drying areas have also been undertaken with TELEMAC-3D. This code solves the Navier-Stokes equations with a free surface, assuming a hydrostatic pressure. These equations are sometimes called 3D Shallow water equations. In the numerical procedure, the free surface has to be computed with the 2D Shallow water equations and thus routines of TELEMAC-2D are called. The meshes in 3D are made of superposition on the vertical of 2D meshes. With 2D triangles, the 3D elements are prisms. An additional problem in 3D is the fact that when there is no water, the prisms degenerate. To cope with this situation, the drying zones are (formally as explained before) removed from the computational domains, and the same option has to be used in the 2D routines.

Figure 5 shows a velocity field of a 3D computation with tidal flats. It is the same case as the one shown on Figures 3 and 4. Here only the velocity of the free surface has been represented.

TURBULENCE AND BOTTOM FRICTION:

After a dam break, propagation of a moving front on a dry area or in a river depends greatly on turbulence and bottom friction. Reproducing turbulence is another unsolved problem and 2D turbulence models are often criticised. For example the depth-averaged k-epsilon model seems to be well suited for lateral jets, but not for recirculations. Moreover, from a theoretical point of view, a 2D treatment of turbulence is not very satisfying and it is highly probable that only 3D models will give a substantial improvement in this field.

It has been shown (in the case of river Culm for instance) that the choice of the bottom friction has a tremendous effect and induces variations of up to 25% on the flood propagation. 1D identification of the bottom friction is now a standard procedure at LNH and a thesis is underway to extend it to 2D domains. Parameter estimation is however not fitted to dam break simulation, where the friction coefficients are obtained by a simple survey of the would be inundated zones, and this remains a major source of inaccuracy.

CONCLUSION:

It appears that active research on specific numerical techniques closely linked with physics is still necessary for treating propagation over initially dry domains. Satisfactory techniques have been found for dealing with most cases but it cannot be claimed that a definitive solution has been found. Collaboration on this topic, with comparisons with analytical solutions and well documented measures, along with the aim of merging the best ideas stemming from different approaches, should be a commitment.

BIBLIOGRAPHY:

[1] J.-C. Galland, N. Goutal, J.-M. Hervouet, A new numerical model for solving shallow water equations.
Advances in Water Resources. Vol. 14 n°3 June 1991 pp.138-148.
[2] J.-M. Hervouet : TELEMAC, a fully vectorised finite element software for shallow water equations. Computer Methods and Water Resources.
Rabat, Morocco.7-11 Oct. 1991.
[3] J.-M. Hervouet : Element by Element methods for solving shallow water equations with F.E.M.". IX International Conference on Computational Methods in Water Resources. Denver, Colorado,USA.June 9-12,1992.
[4] T.J.R. Hughes, R.M. Ferencz, J.O. Hallquist, Large-scale vectorized implicit calculations in solid mechanics on a CRAY X-MP/48 utilising EBE preconditioned conjugate gradients.
Computer Methods in Applied Mechanics and engineering 61(1987) 215-248.
[5] N. Goutal : Finite element solution for the transcritical shallow-water equation.
Mathematical Methods in the Applied Sciences. Vol.11,503-524.1989.
[6] A.N. Brooks,T.J.R. Hugues : Streamline Upwind Petrov Galerkin for convection dominated flows with particular emphasis on the Navier-Stokes equation.
Computer Methods in Applied Mechanics and Engineering. 32(1982) 199-259.

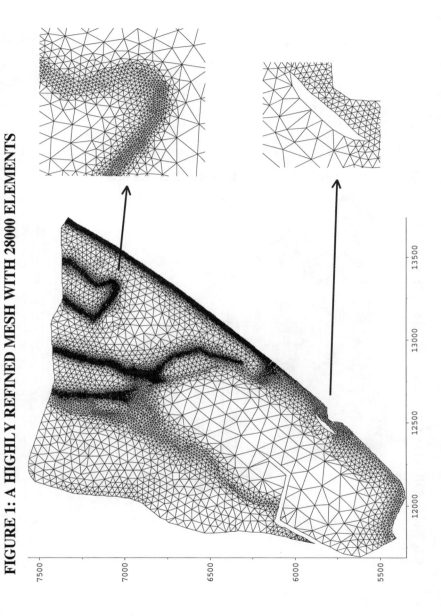

FIGURE 1: A HIGHLY REFINED MESH WITH 28000 ELEMENTS

FIGURE 2 : INUNDATION EXTENT AFTER A DYKE BREAK

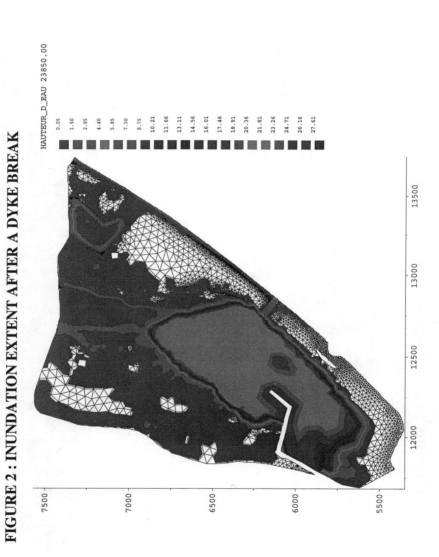

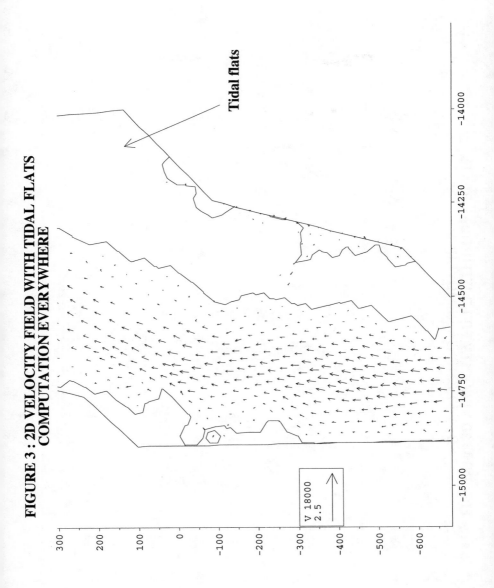

FIGURE 3 : 2D VELOCITY FIELD WITH TIDAL FLATS COMPUTATION EVERYWHERE

Tidal flats

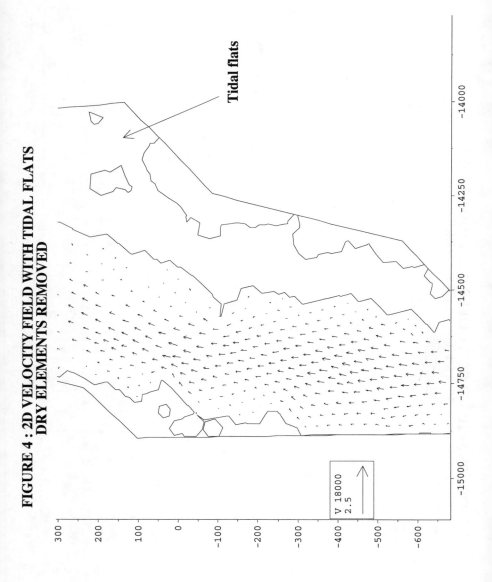

FIGURE 4 : 2D VELOCITY FIELD WITH TIDAL FLATS DRY ELEMENTS REMOVED

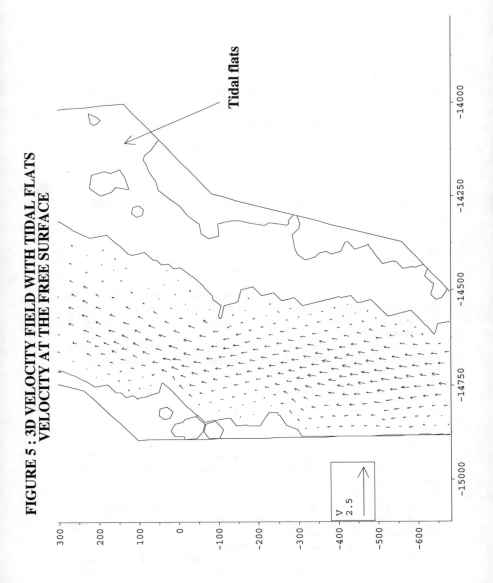

FIGURE 5 : 3D VELOCITY FIELD WITH TIDAL FLATS
VELOCITY AT THE FREE SURFACE

Tidal flats

Adaptive Numerical Methods
for Transport Dominated Processes

Alfio Quarteroni

Department of Mathematics, Polytechnic of Milan, and CRS4, Cagliari

Abstract

The numerical approximation of processes of transport and diffusion can be very challenging when the diffusion is small compared to the convection.

Standard Galerkin methods need to be stabilized in order to eliminate spurious oscillations.

We review some techniques that have been proposed in recent years which eliminate numerical oscillations without downgrading the accuracy of the numerical solution.

1. Introduction

The purpose of this note is to address some recent ideas in the approximation of advection-diffusion processes. It is well known that when the advection domin ates the diffusion, standard numerical methods may provide poorly accurate solutions. Even worse, often the latter are affected by spurious oscillations that arise from unresolved boundary layers.

In a context of finite difference (or finite volume) approximation, the remedy is to resort to one-sided (or upwinded) schemes, although this may lead to low order methods.

When adopting finite element approximations, the same kind of stabilization can be achieved by switching from Galerkin to a Petrov-Galerkin approach. In this case a certain analogy with the finite difference method can be recovered by noticing that in the Galerkin approach the spaces of shape

Work partially supported by Fondi M.U.R.S.T. 40% and by Sardinian Regional Authorities.

and test functions are the same whereas they differ in the Petrov-Galerkin case, breaking therefore the "symmetry" of the choice.

In the following pages we will recall the Galerkin approximation, then we will introduce the stabilized Petrov-Galerkin methods (including the widely used SUPG and GaLS methods) and make comments on these approaches. Next we will address the issue of heterogeneous domain decomposition methods and some recent generalizations that look very well suited for resolving steep gradient layers in advection-diffusion equations. Almost invariably, stability is achieved by adapting the method to the direction of the convective field.

Finally, we will mention the extent at which these equations can be useful to solve more complex flow problems.

2. Advection-diffusion equations and Galerkin approximation

A paradygm for advection-diffusion problems with dominating advection is provided by the following steady boundary-value problem:

$$(2.1) \quad \begin{cases} -\varepsilon \Delta u + \operatorname{div}(\mathbf{b}u) = f & \text{in } \Omega \\ u = 0 & \text{on } \partial\Omega \end{cases}$$

where: Ω is a bounded two-dimensional domain, $\partial\Omega$ denotes its boundary, $\varepsilon > 0$ is a small quantity (the viscosity coefficient), $\mathbf{b} = (b_1, b_2)$ is the convective field and f is a source term.

Both f and $\mathbf{b}$ are prescribed. We may assume that $|\mathbf{b}| \simeq 1$.

It is well known that if ε is much smaller then 1, then the solution of (2.1) may exhibit boundary-layers as it must attain the Dirichlet value $u = 0$ at $\partial\Omega$.

The weak (or variational) formulation of (2.1) reads:

$$(2.2) \quad \begin{cases} \text{find } u \in H_0^1(\Omega) \text{ such that} \\ \displaystyle\int_\Omega \varepsilon \nabla u \cdot \nabla v \, dx - \int_\Omega (\mathbf{b}u) \nabla v \, dx = \int_\Omega f v \, dx \quad \forall\, v \in H_0^1(\Omega) \end{cases}$$

where we have defined the space of admissible solutions as:

$$H_0^1(\Omega) = \{ v \in L^2(\Omega) : \nabla v \in [L^2(\Omega)]^2 \, , \, v = 0 \text{ on } \partial\Omega \}$$

and $L^2(\Omega)$ is the space of measurable functions that are squared integrable in Ω.

The norms of $L^2(\Omega)$ and $H_0^1(\Omega)$ can be defined respectively as follows

$$\|v\|_0 = \left(\int_\Omega |v(x)|^2 dx \right)^{1/2} \ ,$$

$$\|v\|_1 = \left(\int_\Omega |\nabla v(x)|^2 dx \right)^{1/2} \ .$$

For the definition of these spaces and the derivation of the formulation (2.2) see, e.g., Lions and Magenes (1972).

The Galerkin approximation of (2.2) is defined as follows.

Let $\{V_h, h > 0\}$ be a family of finite dimensional subspaces of $H_0^1(\Omega)$. For instance, V_h can be the space of continuous functions that are polynomials of linear (or quadratic) degree over each triangle K of a triangulation T_h of the domain Ω. More precisely, we have

$$\overline{\Omega} = \bigcup \{K, K \in T_h\} \text{ and}$$

$$V_h = \{v_h \in C^0(\overline{\Omega}) : v_{h|K} \in \mathbb{P}_r(K), v_{h|\partial\Omega} = 0\}$$

where $\mathbb{P}_r(K)$ is the space of polynomials in K whose global degree is not greater than r.

In such a case V_h is called space of finite elements: linear if $r = 1$, quadratic if $r = 2$. The parameter h denotes the maximum diameter of the triangles K.

The Galerkin approximation to (2.2) is provided by the following finite dimensional problem:

(2.3)
$$\begin{cases} \text{find } u_h \in V_h \text{ such that} \\ \displaystyle\int_\Omega \varepsilon \nabla u_h \cdot \nabla v_h dx - \int_\Omega (bu_h) \cdot \nabla v_h dx = \int_\Omega f v_h dx \quad \forall \ v_h \in V_h \end{cases} .$$

In order to simplify the notation, let us set

$$a(u, v) = \int_\Omega \varepsilon \nabla u \cdot \nabla v dx - \int_\Omega (bu) \cdot \nabla v$$

so that the Galerkin solution u_h satisfies

(2.4) $$u_h \in V_h : a(u_h, v_h) = \int_\Omega f v_h dx \quad \forall \ v_h \in V_h.$$

It is well known that, if h is not small enough with respect to $\varepsilon/|b|$, the Galerkin solution will exhibit oscillations near the layers (see, e.g., Quarteroni and Valli (1994) for a discussion on this issue). Figure 1 shows the result obtained for problem (2.1) when $\Omega = (0,1)^2$, $b = (1,2)^r$, $\varepsilon = 10^{-2}$ and boundary values prescribed on u are those outlined in the picture. Piecewise linear finite elements are used, with a uniform grid corresponding to $h = 1/20$.

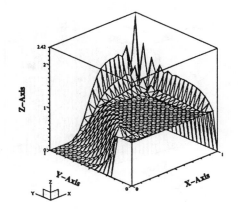

Fig. 1. Galerkin solution

The classical artificial diffusion method consists of generating a Galerkin approximation to a perturbed problem of the form (2.1) in which ε is replaced by $\varepsilon + H$, where H is proportional to h.

This amounts to solve the new problem:

(2.5) $\qquad$ find $u_h^{art} \in V_h : a_h(u_h^{art}, v_h) = \int_\Omega f v_h dx \quad \forall \, v_h \in V_h$

where

$$a_h(u, v) = a(u, v) + H \int_\Omega \nabla u \cdot \nabla v dx.$$

In this case u_h^{art} is free of oscillations no matter how large is h. However, the accuracy of this approximation is compromised by the presence of the stabilizing term containing H. Indeed, one has in this case that

$$\sup_{\substack{v_h \in V_h \\ v_h \neq 0}} \frac{|a_h(u, v_h) - \int_\Omega f v_h|}{\|v_h\|_1} = O(h)$$

where u is the exact solution to (2.1). This means that the scheme (2.5) is *consistent*, but only first order accurate with respect to h, no matter how large the polynomial degree r of the finite element solution is.

See Figure 2, which reports the solution u_h^{art} corresponding to the same problem as in Figure 1, with the only difference that now $\varepsilon = 10^{-4}$.

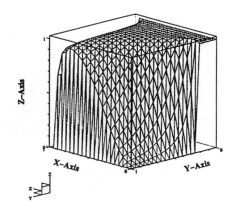

Fig. 2. The Galerkin solution with artificial diffusion.

3. Stabilized Galerkin Methods

The methods that we consider in this Section have the advantage of being stable and generate an artificial diffusion that pollutes the approximate solution only parallel to the streamlines, and not everywhere as the method (2.5) does.

In its most general form, the stabilized method that we consider reads

(3.1) $$\text{find } u_h^s \in V_h : A_h(u_h^s, v_h) = F_h(v_h) \quad \forall\, v_h \in V_h$$

where A_h and F_h need to be specified. The desirable property enjoyed by any such method is that the exact solution u of problem (2.1) satisfies:

(3.2) $$A_h(u, v_h) - F_h(v_h) = 0 \quad \forall\, v_h \in V_h.$$

This property of *strong consistency* (which was not shared by the method (2.5)) allows the solution u_h^s of (3.1) to approximate u with the optimal order of accuracy (super linear when using quadratic elements).

If we denote by $Lw = -\varepsilon\Delta w + \operatorname{div}(\mathbf{b}w)$ the advection-diffusion operator introduced in (2.1), we can split L into $L_S + L_{SS}$, where L_S and L_{SS}, denote its symmetric and skew-symmetric part respectively. We have:

(3.3)
$$L_S w = -\varepsilon\Delta w + \frac{1}{2}w\operatorname{div}\mathbf{b}$$
$$L_{SS} w = \frac{1}{2}[\operatorname{div}(\mathbf{b}w) + \mathbf{b}\cdot\nabla w].$$

The reason behind such terminology is that

$$(L_S w, v) = (w, L_S v) \quad , \quad (L_{SS} w, v) = -(w, L_{SS} v)$$

for all $w, v \in H_0^1(\Omega)$. Here we have denoted by $(\varphi, \psi) = \int_\Omega \varphi \psi dx$ the scalar product of $L^2(\Omega)$.

Then we define

(3.4) $$A_h(w, v) = a(w, v) + b_h(w, v),$$

(3.5) $$F_h(v) = (f, v) + c_h(v),$$

where

(3.6) $$b_h(w, v) = \sum_{K \in T_h} \delta \int_K Lw \frac{h_K}{|\mathbf{b}|} (L_{SS} + \rho L_S) v dx$$

and

(3.7) $$c_h(v) = \sum_{K \in T_h} \delta \int_K f \frac{h_K}{|\mathbf{b}|} (L_{SS} + \rho L_S) v dx.$$

Symbol explanation is as follows: for all finite element K, h_K denotes its diameter, ρ may range from -1 to 1, and $\delta > 0$ is a stabilization parameter that can be chosen suitably.

The most remarkable cases are those corresponding to $\rho = 0$ (the SUPG, Streamline-Upwind-Petrov-Galerkin method, see Hughes and Brook (1982) and also Johnson (1987)), $\rho = 1$ (the GaLS, Galerkin-Least-Squares method, proposed by Hughes, Franca and Hulbert (1989)) and $\rho = -1$ (the DWG, Douglas-Wang-Galerkin method, introduced by Franca, Frey and Hughes (1992) extending to advection-diffusion equations an idea used by Douglas and Wang for the Stokes equations).

For the reader's convenience, let us formulate in details the GaLS method. Starting from (3.1) and using (3.4)-(3.5) and (3.6)-(3.7) with $\rho = 1$ we easily obtain:

(3.8)
$$u_h^s \in V_h : a(u_h^s, v_h) + \sum_{K \in T_h} \delta \int_K \left[\rho(-\varepsilon \Delta u_h^s + \operatorname{div}(\mathbf{b} u_h^s) - f) \frac{h_K}{|\mathbf{b}|} \cdot \right.$$
$$\left. \cdot (-\varepsilon \Delta v_h + \operatorname{div}(\mathbf{b} v_h)) \right] dx = \int_\Omega f v_h dx \quad \forall v_h \in V_h$$

When $v_h = u_h$, the resulting equation differs from the one that we would obtain from the pure Galerkin method due to the presence of the stabilizing term

$$\sum_{K \in T_h} \delta \int_K \frac{h_K}{|\mathbf{b}|} (L u_h^s)(L u_h^s - f) dx.$$

The presence of the square of the operator computed on the solution u_h^s gives the name to the method.

In Figure 3 we report the solution for the same problem considered in Figure 2 using now the GaLS solution for piecewise parabolic elements ($r = 2$).

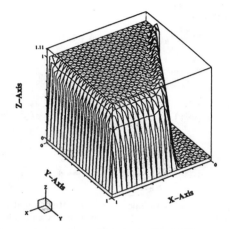

Fig. 3. The stabilized solution obtained by the GaLS method

Remark 3.1 The stabilization procedure defined in (3.1), (3.4)-(3.7) is very general and is not limited to the simple advection-diffusion operator $Lu = -\varepsilon\Delta u + \mathrm{div}(\mathbf{b}u)$ introduced in (2.1). As a matter of fact, if we would replace L by the more general elliptic operator

$$(3.9) \qquad Lu = -\sum_{i,j=1}^{2} \frac{\partial}{\partial x_i}\left(a_{ij}\frac{\partial}{\partial x_j}\right)u + \sum_{i=1}^{2}\frac{\partial}{\partial x_i}(b_i u) + a_0 u$$

where a_{ij}, b_i and a_0 are suitable coefficients, then we have $L = L_S + L_{SS}$ where now

$$L_S w = -\sum_{i,j=1}^{2}\frac{\partial}{\partial x_i}\left(a_{ij}^S\frac{\partial}{\partial x_j}\right)w + \frac{1}{2}(\mathrm{div}\,\mathbf{b} + a_0)w$$

$$L_{SS}w = -\sum_{i,j=1}^{2}\frac{\partial}{\partial x_i}\left(a_{ij}^{SS}\frac{\partial}{\partial x_j}\right)w + \frac{1}{2}[\mathrm{div}(\mathbf{b}w) + \mathbf{b}\cdot\nabla w]$$

and

$$a_{ij}^S = (a_{ij} + a_{ji})/2 \ , \quad a_{ij}^{SS} = (a_{ij} - a_{ji})/2 \ , \quad i,j = 1,2 \ .$$

Then the stabilized Galerkin approximation reads like (3.1) where the definitions (3.4)-(3.7) are still valid, and this time

$$a(u,v) = \sum_{i,j=1}^{2} \int_{\Omega} a_{ij} \frac{\partial u}{\partial x_j} \frac{\partial v}{\partial x_i} dx - \sum_{i=1}^{2} \int_{\Omega} b_i u \frac{\partial v}{\partial x_i} dx + \int_{\Omega} a_0 u v dx$$

is the new bilinear form associated to the general advection-diffusion operator (3.9).

Remark 3.2 At some extent, the stabilized Galerkin method illustrated above can be regarded as a pure Galerkin method with a projection space V_h^b different than the usual one V_h. In the finite element context, the new space V_h^b is obtained by adding to all shape functions of V_h the *bubble-functions*, i.e. a set of cubic algebraic polynomials having support in one single element.
In other words, $V_h^b = V_h \oplus B$, where

$$B = \{v_b \in H_0^1(\Omega) : v_{b|K} = c\lambda_1\lambda_2\lambda_3 , \ c \in \mathbb{R}\}$$

and $\{\lambda_i\}$ are the baricentric coordinates of K.
Then the new Galerkin method reads

(3.10) $$\begin{cases} \text{find}(u_h + u_b) \in V_h^b \text{ such that} \\ a(u_h + u_b, v_h + v_b) = (f, v_h + v_b) \quad \forall\, (v_h + v_b) \in V_h^b \end{cases}$$

It is not difficult to show that (3.10) is indeed a slight perturbation of the classical Galerkin method (2.3). See Brezzi et al. (1992).

Owing to the fulfillment of the strong consistency property, the stabilized solution u_h^s of (3.1) converges to the exact solution u of (2.1) with the good (almost optimal) rate. For instance, if we assume that V_h is a space of finite elements of degree r, then

$$||u - u_h^s|| = O(h^{r+1/2})$$

provided u has derivatives bounded up to the order $r + 1$, when we have set:

$$||v||^2 = \int_{\Omega} (\varepsilon|\nabla v|^2 + |\mu v|^2) dx + \sum_{K \in \mathcal{T}_h} \int_K \frac{h_K}{|\mathbf{b}|} |Lv|^2 dx$$

and $\mu(x) = \operatorname{div} \mathbf{b}(x)/2$.
An extensive analysis of stabilization methods for convection-dominated flows is presented in Quarteroni and Valli (1994), Chapter 8.

4. Stabilization via domain decomposition

When the location of boundary (or internal) layers is known, a very effective approximation method is the one based on the so-called heterogeneous coupling.

With the aim of simplifying our exposition, let us assume that Ω is a domain partitioned in two subdomains, Ω_1 and Ω_2, and that the layer is embedded into Ω_2.

Then instead of solving (2.1) one can solve the *reduced problem*:

$$(4.1) \qquad -\varepsilon \Delta u_2 + \mathrm{div}(\mathbf{b} u_2) = f \quad \text{in } \Omega_2$$

and

$$(4.2) \qquad \mathrm{div}(\mathbf{b} u_1) = f \quad \text{in } \Omega_1.$$

If Γ denotes the common interface of Ω_1 and Ω_2, the two functions u_1 and u_2 need to satisfy on Γ the following matching conditions:

$$(4.3) \qquad -\varepsilon \frac{\partial u_2}{\partial n} + (\mathbf{b} \cdot \mathbf{n}) u_2 = (\mathbf{b} \cdot \mathbf{n}) u_1 \text{ on } \Gamma$$

(continuity of fluxes) and

$$(4.4) \qquad u_1 = u_2 \text{ on } \Gamma_{in}$$

(continuity of solutions), where $\Gamma_{in} = \{x \in \Gamma : \mathbf{b}(x) \cdot \mathbf{n}(x) < 0\}$ and $\mathbf{n}(x)$ denotes the unit normal vector at the point x of Γ directed from Ω_1 to Ω_2.

It can be shown that the solution $\{u_1, u_2\}$ of the reduced problem differs from the restriction $\{u_{|\Omega_1}, u_{|\Omega_2}\}$ of the solution of the original problem by a term that is proportional to ε. Therefore, for high speed flows, solving the reduced problem would provide very accurate approximation to the solution u of the advection-diffusion problem.

In turn, the reduced problem can be solved by an iterative procedure that alternates the solution of the following problems:

STEP 1: problem (4.1) with the Robin condition (4.3) at the interface and $u_2 = 0$ on $\partial\Omega_2 \cap \partial\Omega$;

STEP 2: problem (4.2) with inflow condition (4.4) on Γ_{in} and $u_1 = 0$ on $\Gamma_{in}^1 = \{x \in \partial\Omega_1 \cap \partial\Omega : \mathbf{b}(x) \cdot \mathbf{n}(x) < 0\}$.

In turn, both problems (4.1) and (4.2) will be faced by a suitable method of approximation, e.g., by a Galerkin finite element method.

Clearly, the information on Γ for both step 1 and 2 (i.e., the right hand sides of (4.3) and (4.4)) are provided from the previous iteration.

Since the domain Ω_2 can be "arbitrarily" thin (it should only contain the whole layer), the steep gradient of the solution can be easily resolved by a low number of gridpoints using centered approximations (such as, e.g., pure Galerkin methods).

On the other hand, in the region Ω_1, which may be almost as large as the whole computational domain Ω, the solution u_1 is typically a very smooth one (with low gradient) and can therefore be approximated by any reasonable method suitable for dealing with smooth hyperbolic equations.

This approach has been extensively pursued for high speed flows in recent years. The interested reader may refer to Quarteroni, Pasquarelli and Valli (1992) and Frati, Pasquarelli and Quarteroni (1993).

In Figure 4 we report the solution obtained by the heterogeneous method when each subproblem is solved by the Spectral Collocation Method. The boundary-value problem is still the same but now $\Omega = (-.5, .5)^2$, $\varepsilon = 10^{-3}$, $\mathbf{b} = (-x, 2y)^T$,

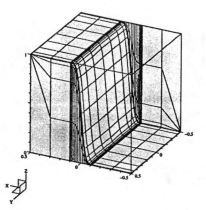

Fig. 4. The numerical solution obtained by the heterogeneous domain decomposition method.

Another approach consists in solving the convection-diffusion method in both domains, i.e.

$$(4.5) \qquad -\varepsilon \Delta u_1 + \operatorname{div}(\mathbf{b} u_1) = f \quad \text{in } \Omega_1 \ ,$$

$$(4.6) \qquad -\varepsilon \Delta u_2 + \operatorname{div}(\mathbf{b} u_2) = f \quad \text{in } \Omega_2 \ ,$$

and the following matching conditions:

$$(4.7) \qquad \varepsilon \frac{\partial u_1}{\partial n} = \varepsilon \frac{\partial u_2}{\partial n} \quad \text{on } \Gamma$$

$$(4.8) \qquad \varepsilon \frac{\partial u_2}{\partial n} - \alpha u_2 = \varepsilon \frac{\partial u_1}{\partial n} - \alpha u_1 \quad \text{on } \Gamma$$

where α is a suitable function (e.g., $\alpha(x) = \mathbf{b} \cdot \mathbf{n}(x)$).

Notice that we are enforcing both the continuity of the solution and the one of its normal derivative on Γ.

When solving (4.5)-(4.8) we apply an iterative procedure that split the problem in Ω_1 from the one in Ω_2 and the corresponding "boundary" conditions at the interface Γ.

Precisely, at the k-th step we solve the following boundary-value problem in Ω_1:

$$(4.9) \quad \begin{cases} -\varepsilon \Delta u_1^k + \operatorname{div}(\mathbf{b} u_1^k) = f & \text{in } \Omega_1 \\[2mm] u_1^k = 0 & \text{on } \partial \Omega_1 \setminus \Gamma \\[2mm] \varepsilon \dfrac{\partial u_1^k}{\partial n} = \varepsilon \dfrac{\partial u_2^{k-1}}{\partial n} & \text{on } \Gamma_{in} \\[2mm] \varepsilon \dfrac{\partial u_1^k}{\partial n} - \alpha u_1^k = \varepsilon \dfrac{\partial u_2^{k-1}}{\partial n} - \alpha u_2^{k-1} & \text{on } \Gamma \setminus \Gamma_{in} \end{cases}$$

then in Ω_2 we solve:

$$(4.10) \quad \begin{cases} -\varepsilon \Delta u_2^k + \operatorname{div}(\mathbf{b} u_2^k) = f & \text{in } \Omega_2 \\[2mm] u_2^k = 0 & \text{on } \partial \Omega_2 \setminus \Gamma \\[2mm] \varepsilon \dfrac{\partial u_2^k}{\partial n} - \alpha u_2^k = \\ = \vartheta \left(\varepsilon \frac{\partial u_1^k}{\partial n} - \alpha u_1^k \right) + (1-\vartheta) \left(\varepsilon \frac{\partial u_2^{k-1}}{\partial n} - \alpha u_2^{k-1} \right) & \text{on } \Gamma_{in} \\[2mm] \varepsilon \dfrac{\partial u_2^k}{\partial n} = \varepsilon \dfrac{\partial u_1^k}{\partial n} & \text{on } \Gamma \setminus \Gamma_{in} \end{cases}$$

where $0 < \vartheta < 1$ is a suitable relaxation factor.

Notice that when solving a boundary-value problem in a subdomain Ω_i we enforce the matching Neumann condition (4.7) at the outflow boundary (i.e. where $\mathbf{b} \cdot \mathbf{n}_i \geq 0$) while the matching Robin condition (4.8) is satisfied at the inflow boundary (i.e. where $\mathbf{b} \cdot \mathbf{n}_i < 0$).

Here $\mathbf{n}_i$ denotes the unit normal on $\partial \Omega_i \cap \Gamma$ directed outward for $i = 1, 2$.

This boundary treatment prevents from the generation of spurious internal layers due to the presence of the internal boundary Γ.

The algorithm (4.9)-(4.10) has been given the name of Adaptive Robin-Neumann method (ARN): it has been proposed and investigated in Carlenzoli and Quarteroni (1994).

In Fig. 5 a computational domain is split into two subdomains, and arrows indicate the flow direction across the interface. The letter R placed on a side of the interface indicates that there is a Robin condition enforced on that side, while N stands for Neumann. At right we report the number NIT of subdomain iterations that are requested to the ARN method (with $\vartheta = 0.9$) to converge. We stress the fact that NIT is virtually independent of ε, and this witnesses the robustness of the ARN method to treat high speed flows.

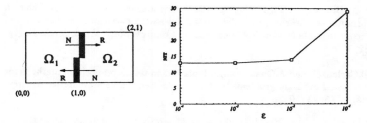

Fig. 5. The computational domain split into two subdomain for the Robin-Neumann domain decomposition method (left) and number NIT of subdomain iterations that are requested to the ARN method to converge (right).

5. Generalizations

Besides its own interest, problem (2.1) can provide a computational kernel of many other complex problems that are currently used in fluid dynamics.

A remarkable instance is provided by the compressible (or incompressible) Navier-Stokes equations, another one by the viscous model of Shallow Water Equations. In both situations, the use of fractional-step schemes for the time-marching can yield a system of equations of convection-diffusion type. See, e.g., Quarteroni and Valli (1994), Chapter 13 and Agoshkov et Al. (1993). An effective solution of this system may ensure the efficiency of the overall numerical procedure.

Acknowledgements

I wish to thank C. Carlenzoli, F. Pasquarelli and F. Saleri for providing the figures of this paper.

References

V.I.Agoshkov, D.Ambrosi, V.Pennati, A.Quarteroni and F.Saleri (1993), Mathematical and Numerical Modelling of Shallow Water Flow, Computational Mechanics 11, pp.280-299.

F. Brezzi, M.-O. Bristeau, L.P. Franca, M. Mallet and G. Rogé (1992), A relationship between stabilized finite element methods and the Galerkin method with bubble functions, Comput. Meth. Appl. Mech. Engrg. 96, 117–129.

A.N. Brooks and T.J.R. Hughes (1982), Streamline Upwind/Petrov-Galerkin formulations for convection dominated flows with particular emphasis on the incompressible Navier-Stokes equations, Comput. Meth. Appl. Mech. Engrg. 32, 199–259.

C.Carlenzoli and A.Quarteroni, Adaptive domain decomposition methods for advection-diffusion problems, Quaderno del Seminario Matematico di Brescia, n. 13/93 (1993).

L.P. Franca, S.L. Frey and T.J.R. Hughes (1992), Stabilized finite element methods, I. Application to the advective-diffusive model, Comput. Meth. Appl. Mech. Engrg. 95, 253–276.

A.Frati, F.Pasquarelli and A.Quarteroni, Spectral approximation to advection-diffusion problems by the fictitious interface method, J. Comput. Phys. 107, 201–212 (1993).

T.J.R. Hughes, L.P. Franca and G.M. Hulbert (1989), A new finite element formulation for computational fluid dynamics: VIII. The Galerkin/Least-Squares method for advective-diffusive equations, Comput. Meth. Appl. Mech. Engrg. 73, 173–189.

A.Quarteroni, F.Pasquarelli and A.Valli, Heterogeneous domain decomposition: principles, algorithms, applications, in *Fifth International Symposium on Domain Decomposition Methods for Partial Differential Equations*, D.E. Keyes, T.F. Chan, G. Meurant, J.S. Scroggs and R.G. Voigt eds., SIAM, Philadelphia, pp. 129–150 (1992).

A.Quarteroni and A.Valli, *Numerical Approximation of Partial Differential Equations*. Springer Verlag, Heildelberg, (1994).

The Numerical Simulation of Wetting and drying Areas using Riemann Solvers

G. W. Tchamen   **R. Kahawita**
Dépatement de génie civil, École Polytechnique de Montréal
C.P. 6079, succ. A, Montréal (Qc), Canada, H3C 3A7

ABSTRACT

The desirable properties of monotonicity and robustness displayed by Riemann solvers applied to homogeneous hyperbolic systems make this class of schemes an appropriate choice for the numerical approximation of the solutions to the Saint-Venant flow equations. The simulation of flooding and drying presents several difficulties. Attempts at addressing the problem by most developers of numerical code have resulted in physically unrealistic behaviour such as negative water depth, failure to conserve mass or "wiggles" which generally lead to computer runtime errors in the form of overflows or attempts to evaluate the square root of a negative quantity. This paper proposes some solutions and methods to circumvent these difficulties and will be discussed briefly. In the case of a horizontal bed, it is shown that Riemann solvers are able to treat areas of flooding and drying without resort to any special artifice. When a varying bathymetry is being considered, some precautions in the treatment of partially flooded cells (i.e. cells containing the front) have to be taken since in these regions, the discretised equivalent of the physical law begins to lose it's validity. The proposed method has been evaluated by comparison with an analytical solution for the dambreak flow along a sloping channel.

1-   INTRODUCTION

In the practical modelling of free surface flows, there are many real situations in which the numerical model must represent unsteady flows over regions which become alternately covered and uncovered during the computation depending on the flow conditions. Common cases of such phenomena include alternate flooding and drying of tidal flats in an estuary, the breach in a dike or the overtopping of a levee, both cases resulting in flooding of a previously dry plain. An extreme example would be the propagation of the surge wave resulting from the catastrophic failure of a dam on a dry valley downstream. The simulation of irrigation flows may also result in the appearance of similar situations where portions of the channel transporting the flow may temporarily dry out due to excessive withdrawal.

It is clear that at a conceptual level, this question addresses a limiting case where the validity of the mathematical model in the physical representation of the advancing or receding front separating wet and dry zones is uncertain. Such a moving boundary problem may be tackled using Lagrangian or Eulerian formulations. Solution techniques belonging to each of these classes have been proposed. However, very few of the proposed algorithms seem to have clearly identified how the failure of the mathematical representation of the physics is transalated at the discrete level, causing trouble in a scheme which may well have provided satisfactory performance for a 'covered field'. A review of previous proposals will show that most developers try to circumvent numerical difficulties in these regions by simply removing computational grids in which the water depth goes below a certain 'critical' value H_{min}. Experience indicates that this threshold value is not only scheme dependent but even problem (flow regime) dependent. This approach results in global mass conservation errors and at a practical level, predicts abnormally large values of velocity near the front. For a given problem, the difficulty lies in the absence of a suitable definition for H_{min}.

The present approach has been developed within the framework of Riemann solvers. It will be shown that using these mathematical tools, which possess a strong physical interpretation associated with them, many interesting points crucial to the understanding of the problem are highlighted. Riemann solvers are capable of simulating the rapidly varied flow that results from the instantaneous failure of a dam on a horizontal channel without the need for any special treatment. The case of a varying bed raises the problem of proper evaluation of the volumetric bathymetry source term for partially wet (or dry) cells containing the front. With this class of solvers, the complicated regions of the problem may be isolated through a correct evaluation of the bathymetry term. The exact integration of this 'slope' term would in general, require subgrid information about the position of the front. To the authors, this option at the present time seems difficult to reconcile with the primary requirement of an efficient general purpose code. Consequently, the solution proposed here is not an exact integration of the source term containing the bathymetry, but an approximation which nevertheless results in a computationally stable representation. The accuracy of such an artifice will be evaluated by comparison with an analytical solution for the dambreak flow on a sloping bed.

2-   REVIEW OF PREVIOUSLY PROPOSED SOLUTIONS

In moving boundary problems of this type, it is natural to consider a grid that follows the front. These Langrangian approaches have been tried by Vasiliev (1970), Akanbi and Katopodes (1988), and Dimonaco and Molinaro (1983). The idea is attractive, but in general fairly difficult to implement in more than one space dimension. Even in one dimension, the correct procedure for updating the front location for the complete Saint-Venant equations including all source terms is not trivial. Akanbi and Katopodes (1983) experimented with an empirical relation for the front velocity which took into account bed resistance. The results presented by these authors were limited to horizontal channels.

The most popular solution is linked to the minimum 'critical' depth concept applied on Eulerian fixed grids. Cells are declared dry and excluded from the computational domain as long as the water depth in the cell is less than the critical value H_{min}. Quite apart from the arbitrariness in the choice of minimum value, this alternative displays the following disadvantages:
- problems with global mass conservation
- neither stability nor convergence may be guaranteed
- appearance of abnormally large velocity values near the front
- a potential retarding of the front velocity when waiting for a downstream dry cell to be considered sufficiently wet, which biases the numerical results.

It is significant that these approaches have never been evaluated by comparison with known solutions. Recently, Meselhe and Holly (1993) revisited the idea of using a modified inertial free equation in the "front" zone. This assumption is probably reasonable for irrigation flows, but too crude for dambreak flows or overland flows over steep slopes. During the preparation of the present communication, the work of Polatera and Sauvaget (1988), in which similar remarks about the need of a subgrid model to properly integrate source terms for a non-horizontal bathymetry was discovered. These authors also felt the necessity of a scheme that mimics the real monotonic properties of flow variables in the critical zone. Since their original scheme did not possess these properties, they were forced to resort to some 'ad hoc' empirical schemes for the small depth zones. The overall approach seems to lack generality, and according to the authors themselves, still suffers from mass conservation problems.

3- THE PROPOSED SOLUTION
3.1: The basic equations

For ease of presentation, we will restrict ourselves to a single space dimension. It is pertinent to mention however, that the approach has already been extended to two-dimensional computations on an unstructured mesh by the authors.

The conservation form of the shallow water wave equations may be written:

$$U_t + F_x(U) = S_o - S_f \tag{3.1}$$

where U designates the vector of conserved variables, $F(U)$ is the inviscid flux of mass, convection and pressure effect, usually called the convective flux, S_f is the friction slope and S_0 represents gravity effects through the bed slope.

$$U = \begin{bmatrix} h \\ uh \end{bmatrix} , \quad F(U) = \begin{bmatrix} uh \\ (u^2) h + g\dfrac{h^2}{2} \end{bmatrix} \qquad (3.2)$$

$$S_o = \begin{bmatrix} 0 \\ gh(-\nabla z) \end{bmatrix} , \quad S_f = \begin{bmatrix} 0 \\ (C_f) uh \end{bmatrix} , \quad C_f = \dfrac{gN^2 |u|}{h^{\frac{4}{3}}} \qquad (3.3)$$

with h and u being the local water depth and mean flow velocity respectively, N is the Manning coefficient and g the gravity.

The computational domain is partitioned into a finite number of control volumes $[x_{i-1/2} , x_{i+1/2}]$ around a nodal position x_i. The position of the right cell face $x_{i+1/2}$ is defined as $(x_i + x_{i+1})/2$. The differential system (3.1) is integrated over each control volume to produce the discrete equivalent of the conservation law:

$$U_i^{n+1} - U_i^n = dt \left[\frac{F_{i-1/2} - F_{i+1/2}}{dx} + (S_o - S_f)_i^{\,n} \right] \qquad (3.4)$$

This integration technique forms the basis of what is known as the finite volume method. The specific differences between various finite volume schemes are the way in which they approximate the interface convective flux $F_{i+1/2} = F(U(x_{i+1/2}, t))$.

3.2: Framework: The Riemann solver for the homogeneous Saint-Venant system

The finite volume technique is nowadays the dominant method used to numerically integrate the hydrodynamic equations of gas flows. These finite volume schemes approximate the integral form of the basic conservation laws. At each time step, they solve for an integral mean of the flow variables in each cell. A discontinuous, piecewise constant distribution of the integral mean values correspond naturally to a class of schemes pioneered by Godunov (1959). To assess the evolution of the piecewise discontinuous data, appeal is made, in gas dynamics, to the Riemann solution of the model shock tube problem which consists of initial data that is piecewise constant and discontinuous. In shallow water wave theory the analogue to the shock tube is the dambreak problem. Riemann solvers in the context of the shallow water wave equations designate the class of schemes, [Godunov (1959), Roe (1981), Osher (1982) and Van Leer (1982)], that may be viewed

as providing exact or approximate solutions to a "local" model dambreak problem. These schemes have several desirable properties:
- they are autmomatically conservative for homogeneous equations (mass) because they use discrete equivalents of the divergence or 'conservation' form of the basic equations. This also provides good shock capturing capabilities with a correct value for the shock speed.
- they are monotone because they treat each nonlinear wave according to its nature, (compression or rarefaction), and consequently they are oscillation free.
- they have an inherent 'upwind' property built in. In the case of supercritical flows this results in the correct direction of propagation of flow properties from the upstream to the downstream state.
These schemes have been presented in detail by Tchamen and Kahawita (1993a,b) and have proven capable of treating the wetting and drying problem on a horizontal channel bed without having to resort to any special device. To the authors' knowledge, together with the implicit formulation of Zhang *et al* (1992), they are the only schemes able to treat the dambreak problem with a downstream dry bed from a 'cold start', without assuming initial critical flow at the dam site as is usually assumed [Bellos and Sakkas (1987), Akanbi and Katopodes (1988)].
The present authors have also shown [Tchamen and Kahawita (1993a)] that the condition for the existence of a solution to the dambreak problem between two neighbouring cells, indexed l and r respectively for the cell at the left and at the right, is:

$$u_l - u_r + 2(c_l + c_r) > 0 \qquad\qquad (3.5)$$

3.3   Treatment of Source Terms

It has been explained by Tchamen and Kahawita (1993a) how a linearisation followed by an implicit treatment of the fricton (source) term is necessary for the stability of the scheme with regard to the resistance.

The framework defined by Riemann solvers and their underlying assumptions make possible an exact integration of the bed slope source term. The representation of the water level by a piecewise constant cell integral mean, allows the volume integral to be transformed into a boundary sum as detailed by Tchamen and Kahawita (1993 b) and has recently been implemented in a 2D code on an unstructured polygonal mesh.

3.4   Genesis of numerical 'small depth' troubles with Riemann solvers

From our experience with these schemes, the mechanism of 'vanishing depth instability' or VDI takes place in the form of an incorrect approximation to the convective flux (dambreak solution) at the boundary between a cell containing the front and it's neighbour immediately upstream; in these two cells the dambreak solution becomes indeterminate due to violation of the existence conditions (3.5). It appears that this behaviour, which manifests itself in a wave front velocity that is too large, originates from an overestimation of the bathymetry term for this cell during some previous time steps. In our case, the formula used to integrate the bed term were only valid for a completely submerged cell, but can falsify the approximation to the mean velocity in a cell containing the front and therefore just partly wet. This remark highlights the mechanism for the occurrence of VDI. An accurate, or at least a well bounded evaluation of the bathymetry term in the partially wet cell is essential for the stability of the numerical algorithm.

3.5   Cures

As a correct evaluation will need a difficult and computationally inefficient subcell resolution to locate the position of the wave front, particularly over a polygonal unstructured mesh, three proposals to control the effect of this slope term (to be implemented in the cell containing the front) are made:
(A) Set velocities to zero in the cell
(B) Remove the bathymetry term
(C) Evaluate the volumetric slope term by the approximate formula

$$\int_{\Omega_i} h \nabla Z \, dv = \overline{\nabla} Z \int_{\Omega_i} h \, dv \qquad (3.6)$$

In the present work, these three alternatives (A), (B) and (C) have been investigated. This corresponds to an alteration (an increase) in the momentum. This modification may be considered as being equivalent to a physical change in the original bathymetry. Consequently, the numerical solution so obtained will not deviate appreciably from the solution to the original problem as long as the modifications made are not important. Method (C) may be a valuable alternative if the approximation made to the mean bottom slope is good. The formula (3.6) would be exact if the slope is constant over the control volume, i.e in the case of a linearly varying bathymetry.

3.6   Remarks

(1) At every time step (dt), the bathymetry term causes a change in the velocity given by $V_b = -g*(dt)*(\Delta z/\Delta x)$.
(2) From remark (1), it is clear that for a channel of constant slope ($\Delta z/\Delta x$), the change in velocity due to the bed slope will remain the same. Consequently, the existence of a constant slope does not modify the homogeneous condition for the existence of the dambreak problem (3.5).
(3) For a concave slope $(\Delta z/\Delta x)_l < (\Delta z/\Delta x)_r$, and the effect of the bed slope is to decrease V_b in the downstream direction. This type of slope ensures the existence condition (3.5).

(4) For a convex slope $(\Delta z/\Delta x)_l > (\Delta z/\Delta x)_r$, and the effect is to increase V_b in the downstream direction. This slope type works against the respect of the condition (3.5).

Remarks (3) and (4) have been proven independently using more rigourous arguments by Vila (1986) whose results may be summarised in the following statements:
(*) For a concave bottom slope, the Riemann problem has a solution for initial data spaced sufficiently close, whereas the existence of a solution cannot be guaranteed for a convex slope. The expression "sufficiently close" implies the numerical form (3.5). This means that computationally, it would be possible to obtain a solution as long as it is ensured that the values in two neighbouring cells are sufficiently close. Since the cell mean values are strongly dependant on the grid and particularly on the grid size, this opens the door to a grid adaptive solution that would reconcile Lagrangian and Eulerian approaches to treat this problem.

4-   ILLUSTRATIVE TEST CASES

The first case is the simulation of the frictionless dam-break flow on a horizontal plane with a dry bed downstream. This problem has a well-known analytical solution due to Ritter. The results obtained with the one dimensional solvers based on the flux approximation due to Godunov (1959), Roe (1981) and Osher (1982) appear to provide a satisfactory approximation to the exact solution as evidenced by an inspection of figure (1).

In the second test, shown in figure (2), the friction term has been added by way of a Chezy coefficient (C=30) to the model. For this case, an "exact" analytical solution is unavailable. However, the results obtained with the above solvers provide good agreement with a perturbation solution proposed by Dressler (1952). The sharp non-physical deviation of the Dressler solution for the water depth near the wave front is due to violation of the basic hypothesis involved. Note that the Ritter solution shown is for a frictionless bed. As expected from intuitive physical arguments, the deviations occur near the wave front at small depths and high velocities where the influence of the friction term is likely to be the most significant. These first two cases prove that these numerical schemes are able to handle a horizontal dry bed with or without the friction effects.

The Dam-break flow along a sloping channel is considered in the third case. The three options proposed to deal with the bathymetry term are evaluated with respect to an analytical solution due to Dressler (1958). It is remarkable that the proposal (A) shows a profile [figure (3c)] that have similarities with the results presented by Kawahara (1986).

The final case concerns the modeling of the hypothetical domain whose irregular bathymetry is shown in figure (4a). An inflow discharge is introduced through the narrow arm at the left. The flow is such that a portion of the bed is uncovered. The velocity field over the partially flooded domain is illustrated in figure (4c).

5-   CONCLUSION

The numerical simulation of flows that alternately flood and dry out grid elements within the computational domain has been investigated in the context of the finite volume formulation. Riemann solvers have been shown to provide stable monotonic solutions to the homogeneous shallow water wave equations. This stability is maintained after addition of the friction source term with an implicit formulation. The correct incorporation of the bed slope is more delicate. Three suggestions that should maintain numerical stability have been proposed in the present paper. The comparison of numerical computations with analytical solutions for dambreak flow along a sloping channel permits an evaluation of the accuracy of either of these alternatives. The most promising proposition (C) suggests a technique of integration for which the bathymetry is supposed to vary linearly in each cell. This seems difficult to reconcile with 2D extensions of Riemann solvers which are locally one dimensional [Tchamen and Kahawita (1993b)]. Research is currently in progress to construct a feasible 2D spatial approximation to effectively include option (C).

REFERENCES

AKANBI, A. A., and KATOPODES, N. D., (1988), "Model for Flood Propagation on Initially Dry Land", Journal of Hydraulic Engineering, Vol. 114, No. 7, pp. 689 to 706

BATES, P. D., and ANDERSON, M. G., " A two-dimensional finite-element model for river flow inundation", Proc. R. Soc. Lond. A (1993), 440, pp. 481 to 491

BELLOS, C. V., and SAKKAS, J. G., (1987), "A 1-D Dam-Break Flood-Wave Propagation on Dry Bed", Journal of Hydraulic Engineering, ASCE, pp. 1510 to 1524, Vol. 113, No. 12, 1987

DI MONACO, and MOLINARO, P., (1984), "Lagrangian Finite Element Model of Dam-Break Wave on Dry Bed Versus Experimental Data", Hydrosoft'84 Hydraulic Engineering Software, pp.2-111 to 2-120, Proceedings of the International Conference, Yugoslavia, September 1984

DI MONACO, and MOLINARO, P., (1988), "A Finite Element Two-Dimensional Model of Free Surface Flows: Verification Against Experimental Data for the Problem of the Emptying of a Reservoir due to Dam-Breaking", Computer Methods and Water Resources: Computational Hydraulics, pp.301 to 312, Proceedings of the International Conference, Morocco, 1988

DRESSLER, R. F., (1958), "Unsteady non-linear waves in sloping channels", Proc. Roy. Soc. London Ser. A, Vol. 247, pp. 186 to 198

DRESSLER, R. F., (1952), "Hydraulic Resistance Effect upon the Dam-break functions", J. Res. Nat. Bur. Stand., Vol. 49.

GODUNOV, S. K., (1959), Mat. Sb., 47, 271 , in Russian

KAWAHARA, M., and UMETSU, T., (1986), "Finite Element Method for Moving Boundary Problems in River Flow", International Journal for Numerical Methods in Fluids, Vol. 6, pp. 365 to 386

MESELHE, E. A., and HOLLY, F. M., (1993), "Simulation of Unsteady Flow in Irrigation Canals with a Dry Bed", Journal of Hydraulic Engineering, Vol. 119, No. 9, pp. 1021-1039

OSHER, S., (1982), "Upwind Difference Schemes For Hyperbolic Systems of Conservation Laws", Mathematics of Computation, 38(158), 339-374.

ROE, P. L., (1981), "Approximate Riemann Solvers, Parameter Vectors and Difference Schemes", J. Comput. Physics, 43, 357-372.

TCHAMEN, G.W., KAHAWITA, R., (1993), ''On the Use of Riemann Solvers for the modeling of shallow water equations,
(a) Part I- One dimensional case'', submitted to 'International Journal for Numerical Methods in Fluids'
(b) Part II- Two dimensional case'', In preparation.

USSEGLIO-POLATERA, J. M., and SAUVAGET, P., (1988), "Dry Bed Depths in 2-D Codes for Coastal and River Engineering", Computer Methods and Water Resources: Computational Hydraulics, pp.415 to 426, Proceedings of the International Conference, Morocco

VAN LEER, B., (1982), "Flux-vector splitting for the Euler Equations", in

Lecture Notes in Physics, Vol. 170 (Springer-Verlag, New-York/Berlin), p. 507.

VASILIEV, O. F. (1970), "Numerical Solution of the Non-linear Problems of Unsteady flows in Open Channels", Proceedings, Second International Conference on Numerical Methods in Fluid Dynamics, Berkeley, California.

VILA, J. P., (1986), "Sur la Théorie et L'Approximation Numérique de Problèmes Hyperboliques non Linéaires: Applications aux Équations de Saint-Venant et à la Modélisation des Avalanches de Neige Dense", Thèse de Doctorat, Université de Paris VI

ZHANG, H., YOUSSEF, H., LONG, N. D., and KAHAWITA, R., (1992), "A 1-D numerical model applied to dam-break flows on dry beds", Journal of Hydraulic Research, pp. 211 to 222, Vol. 30, No. 2

Instantaneous water surface and velocity profiles at dimensionless time 7.8

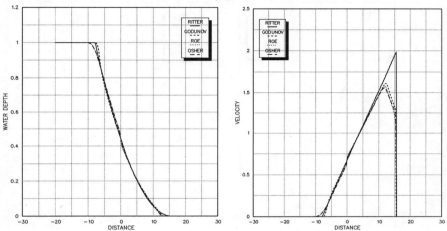

Temporal water surface and velocity graphs (4*Ho measured from dam site)

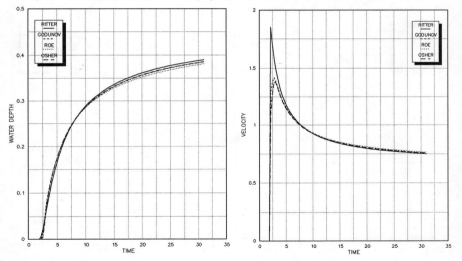

Figure 1: Frictionless Dam—break curves on a horizontal downstream dry bed
(All quantities are dimensionless)

Intantaneous water surface and velocity profiles at dimensionless time 7.8

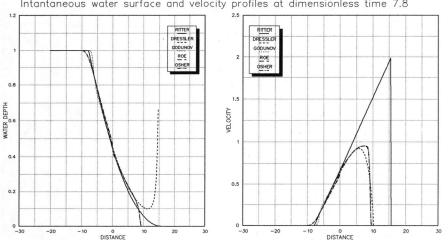

Temporal water surface and velocity graphs (4*Ho measured from dam site)

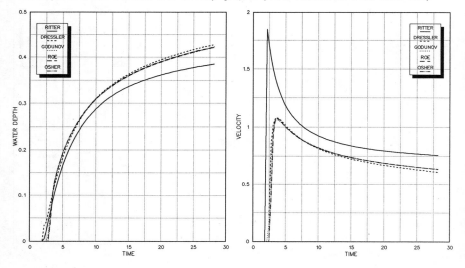

Figure 2: Dam—break curves on downstream dry bed with a Chezy
friction coefficient (All quantities are dimensionless)

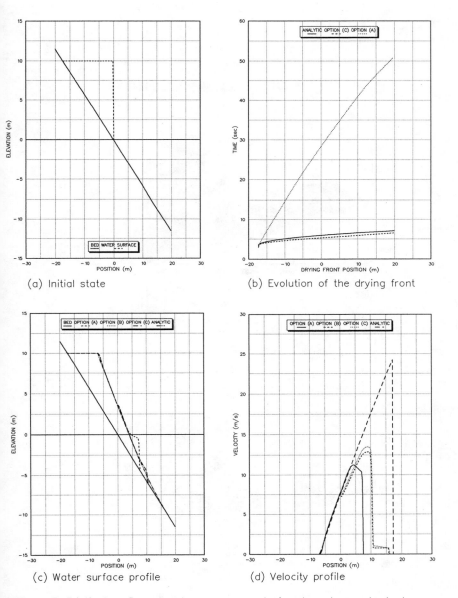

(a) Initial state

(b) Evolution of the drying front

(c) Water surface profile

(d) Velocity profile

Figure 3: Frictionless Dam—break curves on a sloping downstream dry bed

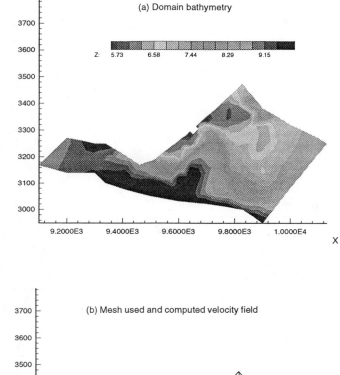

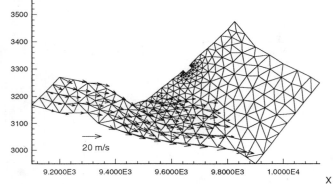

Figure 4: Flood propagation over irregular initially dry areas

Computational and Practical Aspects.

Use of G.I.S. and Other Computer based tools·

Geographical Information Systems and Remote Sensing: a possible support in 2D modelling of flood propagation

F.F. Amighetti, P.P. Congiatu
ISMES S.p.A.
Viale G. Cesare, 29
24100 Bergamo ITALY

A. Di Filippo, P. Molinaro
ENEL-DSR-CRIS
Via Ornato 90/14
20126 Milano ITALY

ABSTRACT

This paper examines some specific aspects regarding the implementation of a 2D mathematical model of flood propagation, namely that of input and output data management. The problems connected with the acquisition, storage and updating of input data, and with the storage, representation and further elaboration of the produced results and information are examined. The use of an integrated system, based on the application of Geographical Information System (GIS) and Remote Sensing (RS) techniques, is discussed. The combination of appropriate remotely sensed images (from airplanes or satellites), available cartographic data and data collected by field surveys, is necessary to create an updated GIS of the study area. The GIS can be used to produce the terrain parameters required by 2D modelling of flood propagation, and to implement a suitable interface between the data and the model. Finally, the GIS can be used as an aid for storage, representation, and analysis of the model output.

INTRODUCTION

The implementation of a 2D mathematical model of flood propagation is particularly complex since a large amount of data on various terrain features is needed. The aim of our work is to implement an integrated system for 2D flood propagation modelling which makes use of different techniques. The mathematical model solving the motion and continuity equations is the core of the system. A Geographical Information System (GIS) of the study area can be used to store the terrain data and to implement the interface between these data and the model. Finally, Remote Sensing (RS) techniques can be used as a tool to retrieve or update the terrain data necessary to the model as shown in Fig. 1.

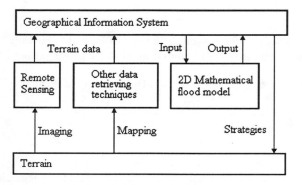

Fig. 1: Integrated system structure

The first section of this paper analyses the type of terrain data required by 2D mathematical modelling of flood propagation. Their enormous amount suggests the use of GIS techniques to simplify management and updating operations. The use of a GIS as an interface between the terrain data and the model can simplify the use of algorithms which compute the various model parameters from terrain data. This can be useful during the model calibration process. Finally, a GIS may assist during the analysis of the time and spatially variable information produced by the mathematical model. The last section of this paper examines the possibility to update cartographic data using RS techniques and limited in situ surveys.

MAIN REQUIREMENTS FOR 2D MODELLING OF FLOOD PROPAGATION

The implementation of a 2D flood wave model typically requires the knowledge of: topographic data of the study area, initial conditions regarding water levels, time evolution of the boundary conditions and a number of land parameters depending on the adopted mathematical solution. Particular care has to be devoted to the management of this large amount of data. An efficient data management strategy has to optimize: the storage procedures, the updating strategies and the creation of an interface between the data and the mathematical model.

For the model calibration process it is useful to employ water levels recorded during previous floods.

A 2D flood model can operate dividing the terrain surface in cells of fixed or different extent. A single cell has its own type of soil, land cover and morphological attributes. A GIS is actually useful if we want an efficient management of all these data and the creation of the model input in a suitable format. In a GIS can also be stored data about local structures like bridges, channels or embankments, usually necessary to simulate the water flow.

A flood propagation model gives essentially as results the water height and the water velocity. These output vary from cell to cell and from time to time. A GIS can be an efficient tool to manage, to visualise and to analyse the large amount of information produced by the model.

An accurate description of the terrain morphology must be based on the availability of a large amount of topographic data. Such information can be supplied directly to the model in the correct format by computing the elevation of each cell through the topographic data stored in the GIS. Additional morphological information about channels, embankments, rivers, etc. is also essential to correctly model the water flow. It is obvious that, for instance, a change of a few decimeters in the evaluation of the height of an embankment can introduce large modifications of the flooded areas in a near flat region.

The land parameters used by the model to describe the terrain features, for example the roughness, have to be computed on the base of the land cover data stored in the GIS. It is important to identify the vegetation cover type, the agricultural lands, the urban land, the snow and ice cover, etc. In urban areas can be useful to define an "urban porosity" coefficient, given by the ratio between the actual built-up area and the total urbanised area (Molinaro, Di Filippo and Ferrari [1]). This coefficient allows to take into account the effective role of buildings on the water storage provided by every cell.

Data about position and features of residential and industrial structures, lines of communication, and other man-made constructions, are also useful to better define the terrain characteristics. This type of data is also important in order to better analyse social and economical damages caused by a possible flood. Finally, soil properties data are useful to take into account the soil/water interaction. For example, data about soil moisture and soil constitution can be important if we want to simulate water infiltration during a flood wave.

GIS TECHNIQUES AS A SUPPORT FOR 2D FLOOD PROPAGATION MODELLING

Nowadays the application of GIS techniques is largely diffused in different fields of Earth sciences. In particular the current evolution towards low cost hardware and friendly software has permitted the implementation of efficient geographical data management strategies in a larger number of research projects.

A GIS is "an organised collection of computer hardware, software, geographic data, and personnel, designed to efficiently capture, store, update, manipulate, analyse and display all forms of geographically referenced information" (ESRI [2]). The geometrical elements usually adopted to describe the geographical elements are: points, lines, and polygons. The geometrical elements describing different features of a region are stored in different GIS layers. Topological relations are established among the elements of a single layer; for example, a polygon contains points and it is bounded by lines. A relation is established among the different layers, on the base of the adopted geographical coordinate system. Finally the geometrical elements can have related numerical and descriptive information managed by a Relational Data Base Management System (RDBMS) (see Fig. 2). The GIS technique is the set of the procedures dedicated to the management of the geographical data. They allow to organise and store the geographical data, to analyse the data, to extract information implementing various algorithms and to display and print a set of selected elements.

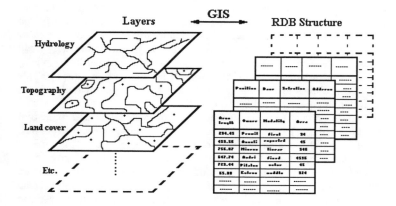

Fig. 2: GIS graphical layers and data structure

The implementation of a GIS can assist 2D flood propagation modelling in different ways. In the GIS all the terrain data necessary to describe the various characteristics of an area can be conveniently organized and stored. In the GIS environment it is possible to implement the algorithms to prepare the input data for a mathematical model in a convenient format. The GIS can be used to store the information produced by the model during a simulation. The GIS also provides tools that can be used during the analysis of the model output.

The retrieval of the necessary terrain data is a complex problem because, even when they are available, they can be of different formats, in different scales, and differently updated. The use of a GIS can be very helpful in merging data coming from different sources and having different formats. Often the existing data are not sufficiently updated. Significant changes can occur in land cover during years. Moreover, the data stored in a GIS become rapidly obsolete, especially in highly urbanized areas. In this case a GIS can be useful to quickly store the retrieved data and, when it is necessary, to update the terrain data using remotely sensed digital images.

The GIS data structure must be designed to store all the data needed to compute the input for flood propagation mathematical models. The data, previously analysed, are those regarding: soil morphology, land cover, and man made structures.

Through digitization of topographic maps: the contour lines, the existing points with elevation attribute and the drainage network can be stored in the GIS. These data can be used to create a raster Digital Elevation Model (DEM). The use of a DEM is an efficient way to supply altimetrical data to the mathematical model of flood propagation. The raster DEM is a regular square grid with an altimetrical attribute at each node. The grid is geographically referenced and covers all the study area. The step of the grid has to be the smallest among those used by the 2D model. A possibility to obtain a DEM directly by satellite or aerial digital images, using digital photogrammetric techniques, is mentioned later in the paper.

Soil cover data are essential. The GIS must store the polygons defining the different land cover classes. These polygons can be obtained digitizing available land cover maps of the area. If no recent map is available, RS techniques may be used. The identification of the following fundamental land cover classes can be useful: water bodies, perennial snow or ice, residential land, industrial land, agricultural land, range land, forest land and rock. Data about seasonal cycle of vegetation cover can be useful especially for the agricultural land class.

Data about the features of different constructions (bridges, embankments, roads, etc.) present on the terrain can be stored in the GIS. As the knowledge of the hydraulic properties of these constructions can be actually important for a given study, all the related data must be elaborated to produce the input for the 2D mathematical model.

Commercial software tools (e.g. ESRI [3]), running on a suitable hardware platform, can be the environment where to set up the GIS.

Usually GIS software works with vector and raster data as an interface between the two types of format. In vector format a line, for example, is represented by a set of pairs of coordinates representing a number of points positioned to reproduce the shape of the line itself (see Fig. 3). Vector format is especially used to store geometrical elements (points, arcs, polygons) optimising the storage memory occupation. Raster format uses a grid of pixels having fixed dimension with a number attributed to every pixel; the coordinates of the objects are defined by their position into the grid (see Fig. 3). Raster format is adopted to store digital images and, for example, the DEM. All the distributed data necessary for the mathematical model can be produced in raster format on a grid determined by the extent of the cells used by the model.

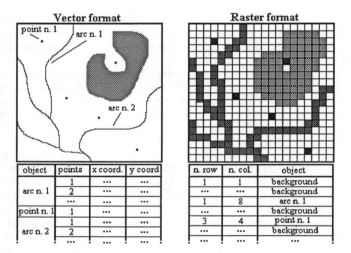

Fig. 3: Vector and raster formats

A change of the grid size or an updating of the input data which bring a variation in the 2D flood model, can be easily managed by means of the GIS. The computational algorithms that represent the interface between the GIS data and the mathematical model can also be stored into the GIS.

The data contained in the GIS can be easily interfaced with the digital images processed by means of an image processing software tool (e.g. ERDAS [4]) that can be adopted to perform the elaboration on the aerial and satellite images necessary in the phase of terrain data updating.

RS TECHNIQUES TO UPDATE INPUT DATA FOR 2D FLOOD PROPAGATION MODELLING

The use of remotely sensed images can be a valuable tool to retrieve the necessary terrain data and correctly implement a 2D flood propagation model.

The locution RS (Remote Sensing) indicates all the procedures relevant to the collection of information about objects with a remotely placed instrument not in contact with the object itself.

RS sensors can operate from different distances: from platforms moving on the earth surface, from towers, from helicopters or airplanes and from artificial satellites orbiting round the earth at different heights. The sensors mounted on satellites and airplanes are particularly useful when the need is to produce terrain data.

The characteristics of a RS system depend both on the sensor design and on the motion of the platform on which it is mounted.

Largely used to study the earth surface are the passive imaging sensors. They produce digital or analogical images of the terrain analysing the reflected solar radiation. They operate by analysing the electromagnetic radiation in different bands into the visible and infrared section of the spectrum.

Active systems, emitting microwave radiation and analysing the component back-scattered by the terrain, are also used to observe and measure natural phenomena.

RS system performance can be evaluated through some parameters. The ground resolution is a measure of the capability to observe small objects. The spectral resolution is the ability to separate objects analysing their aptitude to reflect the different components of the electromagnetic radiation. The radiometric resolution is the ability to distinguish objects emitting, in the same spectral band, radiation with different intensity. Finally, the time resolution is related to the time distance between two consecutive images of the same area produced by the system.

Landsat and SPOT (Systeme Pour l'Observation de la Terre) images are today largely used by the scientific community to produce terrain maps.

The first satellite of the Landsat series (see Fig. 4) was launched in 1972 and placed into a sun-synchronous near polar orbit (any different satellite revolution crosses the equator at the same local time), at the altitude of 900 km (e.g. Lillesand and Kiefer [5]; Elachi [6]).

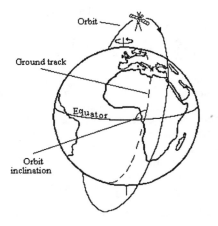

Fig. 4: Sun-synchronous polar orbit of Landsat satellites

The sensor Multispectral Scanner (MSS), carried on board of Landsat satellites, produces images with a ground resolution cell of 80x80 m, analysing solar reflected radiation in 4 spectral bands at a time frequency of 18 days. In 1982 Landsat-4 was launched carrying on board the Thematic Mapper (TM). TM is a multispectral scanner producing images in 6 spectral bands, with a ground resolution cell of 30x30 m. TM also has a thermal band, with a ground resolution cell of 120x120 m, registering the radiation emitted by natural surfaces at ambient temperature. The system Landsat-TM has a time frequency of 16 days. SPOT-1, launched in 1986, carrying on board the Haute Resolution Visible (HRV) sensor, produces images with 3 spectral bands and a ground resolution cell of 20x20 m. The sensor HRV can also produce black and white images with a ground resolution cell of 10x10 m. SPOT-HRV has a time frequency of 26 days but with its capability to produce off-nadir images, the same area can be viewed from different revolutions of the satellite orbit under different inclinations. This opportunity increases the time frequency up to 11 images in the orbital cycle of 26 days and permits the production of stereo views of the terrain. If it is necessary to improve geometrical or time resolution it is possible to use air photographs or digital images produced by airborne scanners. The sensor and platform characteristics have to be carefully considered to select those images that can be useful to collect a particular type of terrain data.

Digital or analogical images produced by satellite or airplane borne sensors can be analysed manually or with computer assisted techniques to extract terrain data that later have to be managed by a GIS. If we want to use these images for land cover classification they have to be corrected geometrically and radiometrically. The correction of images geometrical distortions is necessary to produce a land cover map in an assigned coordinate system. The geometrical correction and the georeferentiation of TM and HRV images can be obtained with standard procedures (e.g. ERDAS [4]) using a set of Ground Control Points (GCP). A GCP is a geographical point of known coordinates with a known position in the image. Working in mountainous areas a more precise geometric correction, including also relief displacement correction, can be performed by means of a DEM, (e.g. Hill and Kohl [7]). The radiometric corrections are connected with sensor calibration, with the geometry of the image recording and with atmospherical variability (e.g. Teillet [8]). Atmospherical effects on remotely sensed images have to be corrected to perform multitemporal studies involving different images of the same area (e.g. De Haan et al. [9]). Many techniques have been examined with regard to this problem to. They are based both on atmospherical sounding data collected during the satellite overflight and on atmospheric mathematical models (e.g. Chavez [10]; Kneyzis [11]; Chavez [12]; Richter [13]; Basu et al. [14]; Hill and Sturm [15]).

The topographical effects on spectral reflectance can be corrected using the data contained in a DEM and this improves the possibility to extract information from the images (e.g. Leprieur et al. [16]; Civco [17]; Janssen et al. [18]). There are many techniques to extract information from digital images; they are all collected under the locution "image processing" (e.g. ERDAS [4]; Jensen [19]). The aim of image processing techniques in RS is the production of clearer terrain images and the automatic objects delineation (e.g. forests, buildings, lakes, etc.).

Digital remotely sensed images can be used for many terrain studies (Table I). Remote sensing techniques, and their capability to produce updated land cover maps, can be really interesting when it is necessary to retrieve the terrain data necessary for 2D flood propagation modelling. By means of Landsat-TM images visual analysis it is possible to extract information about land cover for hydrologic purpouses (Trolier and Philipson [20]). They found particularly convenient to use the bands TM3, TM4 and TM5 in different chromatic compositions. The encouraging result was that "visual analysis of enlarged TM images can provide an accurate and cost effective inventory of hydrologically important land use and cover".

Table I: Principal applications of Landsat TM spectral bands

Band	Limits μm	Band name	Applications
TM1	0.45-0.52	Blue	Water bodies; coastal zone; soil and vegetation; agriculture;
TM2	0.52-0.60	Green	Vegetation; agriculture;
TM3	0.63-0.69	Red	Plant species discrimination; agriculture;
TM4	0.76-0.90	Near-infrared	Vegetation types discrimination; water bodies; soil moisture;
TM5	1.55-1.75	Mid-infrared	Vegetation and soil moisture; snow/clouds differentiation;
TM6	10.4-12.5	Thermal I.R.	Vegetation; soil moisture; thermal mapping;
TM7	2.08-2.35	Mid-infrared	Mineral and rock types; vegetation moisture;

The main results obtained by researchers in analising the different spectral bands to obtain land cover classification suggest an actual capability of RS to map the land cover as needed by a flood propagation mathematical model. It is important to note that the geometric resolution provided by the satellite images usually matches the typical dimension of the elements at present in use for 2D flood wave modelling.

A typical land use/land cover legend is that suggested by Lillesand and Kiefer [5]. Using Landsat-MSS images a coarse distinction of 9 land cover classes can be obtained. These classes are: urban, agricultural land, rangeland, forest land, water, wetland, barren land, tundra, and perennial snow and ice. Using Landsat-TM or SPOT-HRV images, the class named forest land can be subdivided, for example, in: deciduous forest land, evergreen forest land and mixed forest land. The class named rangeland can be subdivided in: herbaceous rangeland, shrub and brush rangeland and mixed rangeland. In a similar way it is possible to work with any other land cover class.

Updating thematic maps using satellite images is a step by step procedure (Mogorovich [21]): selection of images with the needed resolutions, corrections of image distorsions and image classification on the base of data available on the existing maps. The availability of an existing map can be really effective if we want to accelerate the updating procedure. If the existing map is a layer of a GIS, automatic procedures can be activated: training fields are selected to superimpose the obtained images and the existing maps and to extract the new limits of each different land cover class.

Satellite data, as for example SPOT stereo images, can be used to extract a DEM of a mountainous area (e.g. Duperet [22]; Ruesch and Laurore [23]). Photogrammetric techniques can be applied to digital images (produced by airborne or satellite sensors) operating with suitable hardware and software resources.

Finally it can be said that an efficient thematic maps updating procedure can be based on an integrated use of satellite digital images, aerial photography or digital images, existing topographic and thematic maps and in situ surveys, working in a GIS environment and with suitable image processing software tools.

CONCLUSIONS

This paper examines the present feasibility of an integrated system to provide terrain data for 2D flood propagation modelling. The mathematical model is the core of the system while a GIS of the study area can be used to store terrain data and to implement an interface between the data and the model. Finally, RS can be used as a tool to update these terrain data as it is needed by the model.

In RS field ISMES, on behalf of ENEL-CRIS, has experimented different techniques to automatically analyse Landsat-TM and SPOT-HRV images. Satellite images have been used to produce land cover maps, vegetation maps, to evaluate snow pack in alpine regions or to analyse various water bodies feature (Amighetti [24] and [25]). Satellite images have also been used to update topographic maps of the Po river delta region, interfacing RS images and GIS data (Amighetti [26]).

On the base of the described experiences and through an analysis of the terrain data that can be retrieved using RS techniques (Amighetti [27]) we are, at present, working to implement satellite images, airphotographs, and more traditional systems, to create a GIS that can be suitable to easily manage the input and output data of a 2D flood propagation mathematical model.

REFERENCES

[1]   Molinaro, P., Di Filippo, A., Ferrari F. Modelling of flood wave propagation over flat dry areas of complex topography in presence of different infrastructures. International Conference on Modelling of Flood Propagation over Initially Dry Areas, ENEL-CRIS, Milano, Italy, 1994.

[2]   ESRI. Understanding GIS. The ARC/INFO Method. Environmental System Research Institute, Redlands California, 1990.

[3]   ESRI. Arc/Info Users Guide. Environmental System Research Institute, Redlands California, Release Notes 5.0, 1989.

[4]   ERDAS. ERDAS field guide. Earth Resources Data Analysis System, Second edition Version 7.5, Atlanta USA, 1991.

[5]   Lillesand, T.M., R.W. Kiefer. Remote sensing and image interpretation. Second edition by John Wiley and Sons, New York, 1987.

[6]   Elachi, C. Introduction to the physics and techniques of Remote Sensing. John Wiley and Sons, New York, 1987.

[7]   Hill, J., H.G. Kohl. Geometric registration of multi-temporal thematic mapper data over mountainous areas by use of a low resolution digital elevation model. Proc. Earsel Symp., 1988.

[8]   Teillet, P.M. Image correction for radiometric effects in remote sensing. Int. J. Remote Sensing, Vol. 7, No. 12, 1637-1651, 1986.

[9]   De Haan, J.F., J.W. Hovenier, J.M.M. Kokke, H.T.C. Van Stokkom. Removal of atmospheric influences on satellite-borne imagery: a radiative transfer approach. Rem. Sens. of Env., 37, pp. 1-21, 1991.

[10] Chavez, P.S. JR. An improved dark-object subtraction Technique for atmospheric scattering correction of multispectral data. Remote Sensing of Environ. 24:459-479, 1988.

[11] Kneizis, F.X., E.P. Shettle, L.W. Abreu, J.H. Chetwynd, G.P. Anderson, W.O. Gallery, J.E.A. Selby, S.A. Clough. Users guide to LOWTRAN-7. AFGL-TR-88-0177, Air Force Geophysics Laboratory, Bedford, Massachusetts, USA, 1988.

[12] Chavez, P.S. JR. Radiometric calibration of Landsat Thematic Mapper multispectral images. Photogramm. Eng. Remote Sens., Vol. 55, No. 9, pp. 1285-1294, 1989.

[13] Richter, R. A fast atmospheric correction algorithm applied to Landsat TM images. Int. J. of Remote Sensing, Vol. 11, No. 1, 159-166, 1990.

[14] Basu, S., M. Tewari, V.K. Agarwal. A model for retrieval of surface spectral reflectance from satellite radiance measurements using realistic atmospheric aerosols profiles. Int. J. of Remote Sensing, Vol. 11, No. 3, 395-407, 1990.

[15] Hill, J., B. Sturm. Radiometric correction of multitemporal Thematic Mapper data for use in agricultural land-cover classification and vegetation monitoring. Int. J. of Remote Sensing, Vol. 12, No. 7, 1471-1491, 1991.

[16] Leprieur, C.E., J.M. Durand, J.L. Peyron. Influence of topography on forest reflectance using Landsat Thematic Mapper and digital terrain data. Photogramm. Eng. Remote Sens., Vol. 54, No. 4, pp. 491-496, 1988.

[17] Civco, D.L., Topographic normalisation of Landsat Thematic Mapper digital imagery. Photogramm. Eng. Remote Sens., Vol. 55, No. 9, pp. 1303-1309, 1989.

[18] Janssen, L.L.F., M.N. Jaarsma, E.T.M. van der Linden. Integrating topographic data with remote sensing for land-cover classification. Photogramm. Eng. Remote Sens., Vol. 56, No. 11, pp. 1503-1506, 1990.

[19] Jensen, J.R. Introductory digital image processing. Prentice-Hall, 1986.

[20] Trolier, L.J., W.R. Philipson. Visual analysis of Landsat Thematic Mapper images for hydrologic land use and cover. Photogramm. Eng. Remote Sens., Vol. 52, No. 9, pp. 1531-1538, 1986.

[21] Mogorovich, P. Updating thematic maps using TM and SPOT images. RIENA, Space Meeting Proceedings, Rome, 1987.

[22] Duperet, A. Utilisation geomorphometrique d'un modele numerique de terrain calcule par correlation automatique d'images SPOT. Photo Interpretation, No.1990-1, fascic. 4, pp. 31-38, 1990.

[23] Ruesch, S., L. Laurore. Extraction de modeles numerique de terrain calcule par correlation automatique sur des couples d'images stereoscopiques SPOT. Photo Interpretation, No.1990-3 et 4, fascic. 4, pp. 33-36, 1990.

[24] Amighetti, F.F. Elaborazione di immagini da telerilevamento per la sicurezza degli invasi. ISMES report, Doc. RAT-DMM-1090/90, Bergamo, Italy, 1990.

[25] Amighetti, F.F. Elaborazione di immagini da telerilevamento per la sicurezza degli invasi. ISMES report, Doc. RAT-DMM-827/91, Bergamo, Italy, 1992.

[26] Amighetti, F.F. Valutazione delle possibilità di impiego del telerilevamento per l'analisi delle dinamiche costiere in aree prospicienti il delta del Po. ISMES report, Doc. RAT-DMM-55/92, Bergamo, Italy, 1992.

[27] Amighetti, F.F. Applicazioni del telerilevamento da satellite per la produzione di dati sul territorio da gestire con sistemi GIS. ISMES report, Doc. RAT-DMM 1394/93, Bergamo, Italy, 1994.

QUASI-TWO DIMENSIONAL MODELLING OF FLOOD ROUTING IN RIVERS AND FLOOD PLAINS BY MEANS OF STORAGE CELLS

E. Bladé, M. Gómez, J. Dolz.
Hydraulic, Maritime and Environmental Engineering Dept.
Civil Engineering School. UPC.
Gran Capitán s/n. Mod. D-1. 08034 Barcelona SPAIN.

ABSTRACT

A quasi-two dimensional numerical model for flood routing in rivers and flood plains has been developed. In the river, the St Venant equations are integrated using the implicit finite difference Preissmann scheme while the flood plain is modelled by means of storage cells. Mass conservation is imposed at every time step on every cell, using a fully implicit approach between storage cells and the main river.

To avoid instabilities detected when water levels in the river and flood plain are very similar, an automatic time step reduction algorithm is implemented. In this way, the advantages of implicit finite difference methods are maintained. Mass conservation is verified and a sensitivity analysis of the different parameters involved is carried out. Finally, the proposed formulation is compared with other commonly used approaches.

INTRODUCTION

Mediterranean rivers are known for their highly irregular character and periodical hazardous floods. In their lower reaches these rivers usually flow through very extense flat areas, which can be under the level of the river embankments or even under the river bed level. In such cases, when a flood takes place, great extensions of the surrounding plains are inundated. During the flood rise, the water sheet propagates over the initially dry area being driven and limited by natural boundaries such as roads, dikes or high areas. After the flood peak has passed, some parts of this flooded area drain again into the main river, others do it directly into the sea and some lower areas can remain flooded for a long period of time.

In order to quantify the flooded area, different methods of mathematical modelling have been used. The simplest, most commonly used but clearly inadequate in some cases, is the gradually varied flow approach, too conservative and not applicable to large flood plains. The approach of unsteady one-dimensional flow with uniform velocity distributions in the main channel and flood plain is also inadequate in extense flood areas, even considering the flood plain as a storage area with no contribution to the dynamic equation. The previous approaches make the assumption of constant water level in every cross section, which is not true in large flooded areas. To take the effect of different water levels into account, some kind of two-dimensional method is necessary.

Trying to solve accurately the two-dimensional unsteady flow equations in a large area would be extremely expensive in terms of information and computational costs. A more practical method is to use a separate numerical representation of the main channel and the flood plain (eg. Pender [1]). In the present paper the main river is studied with the unsteady one dimensional flow approach, while the flood plain is schematized with what has been called a quasi-two dimensional method: a certain number of intercommunicated storage cells with water level calculated using mass conservation (eg. Cunge [2]).

GOVERNING EQUATIONS

A separate formulation is used in the main river and the flood plain.

Main River

Simulation of flood routing along the river is done through the solution of Saint Venant equations (Cunge [3]). In our case, where there can be lateral inflow, they are:

$$b\frac{\partial y}{\partial t} + \frac{\partial Q}{\partial x} = q \tag{1}$$

$$\frac{\partial Q}{\partial t} + \frac{\partial}{\partial x}\left(\frac{Q^2}{A}\right) + gA\frac{\partial y}{\partial x} - \frac{Q}{A}q = gA[I_0 - I] \tag{2}$$

where y is the flow depth, Q the discharge, x the longitudinal distance, t time, b the top-width of the channel, q the lateral inflow per unit length, A the cross-sectional area of flow, g gravity, I_0 the bed slope and I the friction slope.

Flood plain

Flood plain schematization has been done by means of a number of storage cells or compartments which try to represent the actual topography of the area. The water surface at the whole flood plain is represented by the values of the water levels at every cell. Each cell can be connected to any other cell with a common boundary or to the river. Flood routing in the flood plain is done using only mass conservation. For this purpose it is necessary to consider the volume stored in each cell and its variation due to the discharge between a given cell and the ones surrounding it.

Two different types of links between cells or between the main river and the cells connected to it are possible. The first type is for boundaries formed by dikes, walls, elevated roads or any significant obstacle to the flow that produces a singular head loss and can be represented by a weir type equation. The second, called river type links, are used where there is not an obvious topographic obstacle in the flood plain but it is convenient to divide it into different cells to allow for the different water levels in different parts of the plain. This type of link is schematized with the cross section connecting the two cells and the flow resistance is given by a mean friction coefficient in this section.

a)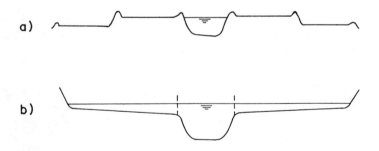

b)

Fig. 1- Links between river and flood plain. Weir (a) and river (b) types.

The mass conservation equation for a given cell (or continuity equation) can be written as

$$A_{sk}\frac{\partial z_k}{\partial t} = \sum_i Q_{ki}(z_k, z_i)$$ (3)

where A_{sk} is the water surface of cell k, z_k the water level at cell k and

Q_{ki} the discharge from cell i to cell k. This equation relates the water level variation of each cell to the total lateral inflow from the main river and neighbouring cells. In the deduction of the continuity equation two hypotheses have been made: 1) the volume stored in one cell is directly related to the water level z_i in that cell; and 2) the discharge between two cells i and k at a given time depends only on the water levels z_i and z_k in those cells.

For weir type links the discharge formula for broad crested weirs is used

$$Q_{ki} = \gamma C_d L (z_k - z_w)^{\frac{3}{2}} \tag{4}$$

where C_d is the water discharge coefficient, L the width of the weir, z_w the weir elevation and γ is a submergence factor which accounts for the influence of the downstream water level (Huber [5]).

For river type links, the discharge between two cells connected by this type of link is obtained with the Manning's formula

$$Q_{ki} = \frac{1}{n} A_{ki} Rh_{ki}^{2/3} I^{1/2} \tag{5}$$

where A_{ki} and Rh_{ki} are the area and the hydraulic radius of the flow cross-section between cells k and i, n is the Manning coefficient and I the water slope between the centers of cells k and i.

MATHEMATICAL METHOD AND MODEL DESCRIPTION

The purpose of the model is to calculate the discharge and flow depth in every section of the river and the water levels in the flood plain at any time during the flood. There are two systems of equations, one for the river and another for the flood plain, linked through the connecting equations that give the discharges between the river and the cells connected to it.

Main river

In the main river the St. Venant equations are integrated using the implicit finite difference Preissmann scheme. With this method, if there are N river sections, to obtain the discharge and flow depth in all of them at a given time, there are $2(N-1)$ equations which result from the discretization of the St. Venant equations, and $2N$ unknowns. The two other equations needed are provided, in subcritical flow, by the upstream and downstream boundary conditions. The boundary conditions of given flow depth or given discharge at the upstream end, and given flow depth or weir type discharge-depth relation at the downstream end have been used.

The system of equations for the river is a set of nonlinear equations which is solved by iterative procedures. Some coefficients, corresponding to the lateral inflow (q on equation (1) and (2)), depend on the water levels of the storage cells connected to the river. When using the Preissmann scheme, the lateral inflow in river section j at time $t_i = i\Delta t$ is $q_j^i = (\sum Q_{jk}^i)/\Delta x$, where Q_{jk}^i is the discharge between cell k and river section j at ime t_i according to the corresponding connecting equation.

The model has the capability of considering that the main river channel separates in two for a certain distance, forming an island between its two branches. In such cases the St. Venant equations are solved in each branch while continuity and energy conservation is imposed at the separation nodes. This causes the matrix of the system of equations to have a bandwidth of 7, instead of 5 as it would be otherwise.

Flood Plain

In the flood plain the water surface is represented by water level z_k at every single cell k. There is the same number of equations as number of cells, and, due to the flow type approach used, there are no spatial partial derivatives. The following time approximation has been used:

$$f = \theta f^{i+1} + (1 - \theta)f^i \tag{6}$$

$$\frac{\partial f}{\partial t} = \frac{f^{i+1} - f^i}{\Delta t} \tag{7}$$

where f^i is the value of any function f at the time t_i and θ is the same temporal weighting coefficient used in the Preissmann scheme ($\theta = 0.6$ for typical applications). This approximation leads to a nonlinear system of algebraic equations with a diagonal system matrix. In these equations the coefficients depend on the values of the discharges between cells and between cells and the river.

To allow for the flood plain draining directly into the sea, it is possible to impose the water level at every given time in a number of cells. This, together with the boundary condition of given flow depths in the river downstream end, enables the model to take into account tidal effects or different sea water levels.

Initial conditions

To integrate the two systems of nonlinear algebraic equations, one for the main channel and the other for the flood plain, it is necessary to know the set of initial values of discharges and flow depths in the river and water levels in the flood plain. For this purpose, the flow depths at every section of the river are calculated with the steady gradually varied flow approach, using the *step method* and a downstream condition of known flow depth.

In the flood plain it is possible to give initial values of water levels at every cell.

RESOLUTION ALGORITHM AND STABILITY

The two systems of equations for the river and the flood plain are nonlinear. This means that the coefficients of the system matrixes depend on the values of the unknowns (discharges and flow depths in the main river, water levels in the flood plain cells). In addition, the coefficients of the matrix corresponding to the river equations depend on the water levels in the cells at time t_i and t_{i+1}, and those of the flood plain equations depend on the discharges and flow depths in the river.

It is possible to solve the two systems of equations in the river and the flood plain in a coupled or an uncoupled way. At an early stage the uncoupled formulation was used. It consisted in solving each system of equations only once in every time increment. The flow depths in the river were calculated using constant flood plain water levels equal to those obtained at the previous time level. After, the cell equations were solved using the river water depths just calculated as constant values. Important oscillations appeared and this uncoupled method proved to be unstable.

A coupled or fully implicit method was chosen. In it, the two systems of equations are solved in a series of iterations, using the values of the unknowns just calculated as approaches to the unknowns for the next calculations.

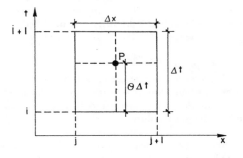

Fig. 2- Finite difference grid.

If the values of the discharges and flow depths in the main river and the values of the water levels in the cells are known at a given time t_i, the unknowns are these values at time t_{i+1}. We will use i for the known time level $t_i = i\Delta t$, and $i + 1$ for the unknown time level $t_{i+1} = (i + 1)\Delta t$.

The resolution method that has been used solves the equations in the main river and the flood plain separately. The algorithm used for calculating the unknowns at time level $i + 1$, once they are known at time level i, is as follows:

1) The discharges and water levels in the river at time level $i + 1$ are calculated by successive approaches. The initial guess for the unknowns at time level $i + 1$ are the discharges and water levels at time level i. During the iterative process water levels in the cells are kept constant and equal to their values at time level i.

2) With flow depths in the river at time level i already known, and using the flow depths calculated in the previous point as approximate values in the river at time level $i + 1$, water levels in the cells at time level $i + 1$ are calculated by successive approaches.

3) The calculations of point 1) are repeated, but using the water levels obtained in point 2) as approximations to water levels in the cells at time $i + 1$, and using the discharges and flow depths calculated in point 1) as initial guess at time level $i + 1$.

In this way, while the iterative process goes on, the last values calculated are always incorporated as approximate values of the unknowns at time level $i + 1$. When the discharges and flow depths in the river are being calculated, water levels in the cells remain constant and vice versa. If in one of these iterations the unknowns are modified in less than a predetermined percentage, convergence has been achieved.

Stability and time step reduction
One of the problems with the storage cells representation is numerical oscillation appearing when water levels are very similar in two connected cells or in the river and the cells connected to it. These oscillations produce instabilities and convergence failure.

In Figure 3 a typical example leading to instability is presented. It is the case of a single storage cell connected to the main channel. When the continuity equation for the cell is being solved, in iteration 1 the new water level is calculated using the flow depth in the river at time level i, an approximation to this flow depth value at time level $i + 1$ and the water level in the cell at time level i. As the initial time step Δt is large, even for small values of the discharge between river and cell a large volume can

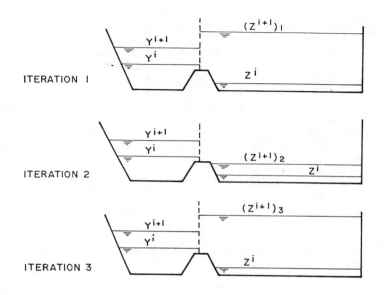

Fig. 3- Convergence failure without time step redution.

be transferred and the water level calculated at time level $i + 1$ $((z^{i+1})_1$ in Figure 3) can be above the river water level. In iteration 2, using water levels at time level i and the water level calculated in iteration 1 as an approximation to water level in the cell at time level $i + 1$, the discharge is in the opposite (cell to river) sense and water level calculated at time level $i + 1$ can be again under the embankment river level. In iteration 3 we are in a similar situation than in iteration 1. Water level calculated in iteration 3 will be the same than that calculated in iteration 1. The results of the following iterations will continue to oscillate and convergence will never be achieved.

Convergence can be assured by reducing the time step, but if it is globally reduced in the whole resolution process, the advantages of an implicit scheme are lost. Pender [1] used an algorithm, for a single cell connected to the river, that halved time step if the water level in the cell rose above the water level at the main channel node connected to it. This does not work with an arbitrary number of cells, as in some of them water level can actually rise above the river water level if those cells are fed by other cells apart from the river.

The present algorithm controls the variations of water levels in two consecutive iterations. If the sum for all the cells of the absolute values of these variations is greater in one iteration than in the previous one, time step is halved, and the last iteration is repeated using the new time step. This new time step will be used until a full initial time step has been completed.

To have an idea of the algorithm efficiency, 13 minutes of computation were needed with a Vax 6420 for an example with the following characteristics: river length: 15 km, flood plain area: 11.25 km^2, study time period: 50 hours, time step: 600 s, number of river sections: 31, number of storage cells: 90, number of weir type links: 174.

MODEL VALIDATION AND SENSITIVITY ANALYSIS

As no real data was available, the model validation could not be fully accomplished. Anyway, some theoretic examples have been studied. Steady states and mass conservation have been verified and a sensitivity analysis has been carried out to know the effects of the different parameters involved in the flow modelling.

First, the good behaviour of the model in the main channel has been verified. The immediate test is proving that the model properly reproduces steady flow. With steady boundary conditions, the difference between the results obtained with the model and those obtained considering steady gradually varied flow are inexistent. Tests have been done with the real topography of the river Ebre between the cities of Tortosa and Amposta, a reach of 15.187 km in the river lower part with a mean slope of 0.04%, represented with 33 sections and with an island represented with three sections for each branch. With this same river geometry and unsteady flow, mass conservation in the river has been verified. Less than 0.03 % of volume is lost in a typical flood. In this test a triangular hydrograph has been used, with discharges varying from 50 to 500 m^3/s for a hidrograph base time of 24 hours and time to peak of 12 hours. Periods of 50 hours have been studied.

Concerning the flood plain modelling, simple examples with a small number of storage cells connected to the river and between them have been studied to test the good behaviour of the model. In Figure 4 an example of a single cell connected to the river through a weir type link is represented. At a first stage water level in the river rises, but there is no flow to the cell as it is still under the weir level. When water depths in the river rise above the weir level, water begins to flow to the cell, where water level quickly increases until it nearly equals the level in the river.

From this moment on the discharge between river and cell is much smaller and water levels increase together in the river and cell. After the peak has passed, water level in the river decreases to the initial value while in the cell it cannot decrease under the weir level and remains there as a stored volume of water.

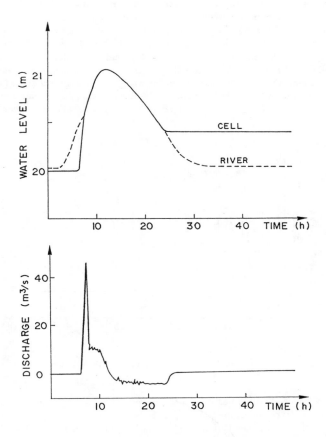

Fig. 4- Single cell connected to the main channel. Water levels and discharges.

Different number of cells and types of links have been used to study mass conservation. Mass errors were always less than 1% with six cells or less, and increased to a maximum of 7% in the previously mentioned case of 90 cells.

When applying the model to a real case, it should be properly calibrated. The parameters that should be calibrated are those coeficients affecting water transfer between cells and cells and river, as well as roughness in the main channel. For this purpose it is important to know how the different parameters involved in the flow modelling affect the results. A sensibility analysis has been carried out in order to show the influence of the following parameters: cells surface, width and weir coefficient for weir type links, roughness (Manning coefficient) and slope in the river type links. What has been studied is the influence of these parameters on the following results and their evolution with time: river discharges (with special attention to the hidrograph at the downstream end), river depths, water levels in the cells and transference discharges between the river and cells.

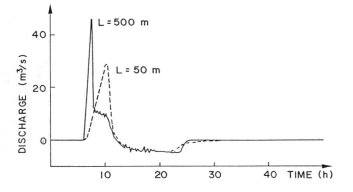

Fig. 5- Effect of weir width in weir type links.

The reference example has been a single cell connected to a section of the river Ebre reach mentioned above, at 1730 m from the downstream end. If the surface of the cell is increased, what happens is that water level in it does not rise so quickly, the difference of water levels in the river and the cell is greater, so is the discharge between them, and the discharge peak takes place later. If the storage cell is big enough, water level in it might never rise over the river embankment. Also, the bigger the cell, the greater hidrograph atenuation in the river is. In weir type links, if the weir width or the weir coefficient are increased, transference discharge increases and water level in the cell equals that in the river at an earlier moment. Peak discharges are greater, they take place earlier (Figure 5) and there is less atenuation in the river.

In river type links, increasing roughness hardly changes the value of the peak discharge through that link, but what actually changes is the instant this peak takes place. Increasing the distance between the cell center and the river (which is the same as decreasing the slope at the transference section) and using transference sections with less area, produce less transference discharges and consequently a similar effect to that of reducing the Manning roughness coefficient.

COMPARISON BETWEEN DIFFERENT APPROACHES

There are different approaches for flood routing in rivers and their flood plains. To prove the advantages of the proposed storage cells method, it has been compared with the more commonly used approaches of steady gradually varied flow and integration of the St. Venant equations. In these other approaches, cross sections which include both the river and the flood plain are used, with only one value for the water level in every cross section and usually also only one velocity value in the whole section. In fact, some models using steady gradually varied flow use different velocity values in the main channel and in the flood plain at each side of it and others that solve the St. Venant equations consider the flood plains to contribute only to the continuity equation (they suppose that there is only storage and no water transport through the flood plain). Anyway, this last assumption hardly affects the results (eg. Gutierrez [7]) and what is significant is that all these other approaches assume that water levels in the river and in the whole flood plain are the same in the whole cross section.

Various theoretical cases with different geometries have been studied using the following methodologies: gradually varied flow, St. Venant equations integration, and storage cells. In such cases a prismatic rectangular channel of 15 km and 0.01% of slope has been used, and flood plains of 0, 250, 500 and 750 m at each side of the river, with 0, 30, 60 and 90 cells respectively. All cells are identical, with a surface of 250000 m^2 (250 × 1000m), and distributed in 0, 1, 2 or 3 rows at both sides of the river, with 15 cells each row, to represent the flood plains of 0, 250, 500 and 750 m respectively (Figure.6). Weir type links have been used between cells and between cells and the river. Cells are interconnected in both transversal and longitudinal directions The output hydrographs and water levels in the river and flood plain obtained with the three metods have been compared. It has been observed that the larger the flood plain is, the more different the results with different methods are.

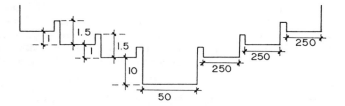

Fig. 6- Flood plains used in the different approaches comparison.

In Figure 7 the results for a 750 m flood plain at each side of the river (90 cells) are represented. When comparing the hidrographs obtained with the storage cells scheme and with the St. Venant equations integration, it can be seen that there is more atenuation with the storage cells method (lower peaks are produced, and later). Figure 7 also shows some problems that appear when integrating the St. Venant equations in the river and the flood plain: when water levels rise above the embankment level, the flood plain is inundated and the wetted perimeter greatly increases; in this way small changes of flow depth and discharge can produce important changes of the friction slope. This makes the numerical method to produce results with discharges that temporarily decrease in the rising stage of the flood (as a result of a sudden rise of the friction slope values), and increase when water levels decrease again under the embankment level. These undesirable consequences of using only a single water velocity value in all the cross section are eliminated with the cells method.

When representing water level evolution in the flood plain, it can be seen that maximum water depths are greater with the cells method, although they are produced later. The flood plain starts to be flooded later but it also remains flooded for a longer period of time. The gradually varied flow approach gives poor information as it cannot reproduce water level evolution with time. In figure 7, water levels in a cell of the third row (the most distant from the main river) at 7 km from the downstream end have been represented.

Maximum water depths are quite similar with the three methods, but they are produced at different time levels with the cells method and the

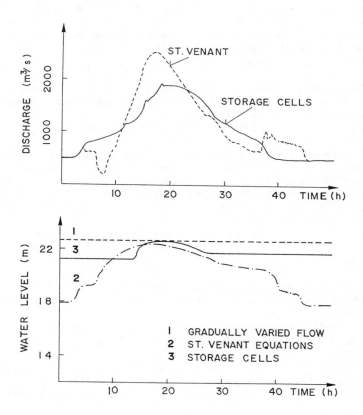

Fig. 7- Comparison of results using different approaches. Downstream output hidrographs and water levels evolution with time in the flood plain.

St. Venant equations integrations, and no time information is possible with the gradually flow approach. The cells method is the only one that can consider different water levels in different points of the same cross section, and water discharges in other directions than the river axis.

CONCLUSIONS

The development of a numerical method to simulate unsteady flood propagation in a combined form in the river and its flood plain is made through a separate schematization. In the river the full St. Venant equations are solved, while the flood plain is represented by means of storage cells where mass conservation is imposed through a fully implicit finite difference scheme.

The automatic time step reduction algorithm that has been implemented, has proven to be a reliable tool which avoids instabilities and maintains the advantages of implicit schemes, which allow working with large time steps, as that reduction is done only when strictly necessary.

Although validation with real data was not carried out, mass conservation is verified and a series of tests proved the good behaviour of the model. Some important features of the model are its capability to consider different water levels in the flood plain and discharges in other directions than the river axis. Consequently, in contrast to other methods, the period of time during which the flood plain remains flooded can be different in different points situated at the same distance from the upstream end.

REFERENCES

1. Pender, G. Maintaining numerical stability of flood plain calculations by time increment splitting. Proc. Inst. Civ. Engr. Wat, Marit & Energy, 1992.
2. Cunge, J.A. Two-dimensional modeling of flood plains, in Mahmood, K, Yevjevich, V. Unsteady Flow in Open Channels. W.P.R. Fort Collins, 1975.
3. Cunge, J.A. Practical aspects of computational river Hydraulics. Pitman, London, 1980.
4. Huber, W.C. EXTRAN (EXtended TRANsport) Users Manual. EPA. Athens, 1987.
5. Abbott, M.B. Computational Hydraulics. Pitman, London, 1979.
6. Henderson, F.M. Open Channel Flow. Macmillan Publishing Co. Inc, New York, 1966.
7. Gutiérrez Hernández, C; Gómez Valentín, M. Modelo de estudio de propagación de avenidas en cauces naturales. Tesina de especialidad. Civil Eng. School, Barcelona, 1991.
8. Bladé i Castellet, E; Gómez Valentín, M. Modelació quasi-bidimensional d'avingudes en un riu i la seva plana d'inundació mitjançant cèl.lules d'emmagatzematge. Tesina d'especialitat. Civil Eng. School, Barcelona, 1993.

Combining a 2-D Flood Plain Model with GIS for Flood Delineation and Damage Assessment

P. Di Giammarco; E. Todini

University of Bologna, Istituto di Costruzioni Idrauliche, Viale del Risorgimento 2; 40136 Bologna, Italy

David Consuegra; F. Joerin; F. Vitalini*

Swiss Federal Institute of Technology; Rural Engineering Department, Soil and Water Management Institute, CH 1015 Lausanne.
(*) On leave from Politecnico di Milano under EC-ERASMUS

ABSTRACT

This paper describes potential applications of Geographical Information System (GIS) for flood hazard delineation and damage assessment. The hydraulic model simulates both 1-D flood routing in the main channels and 2-D propagation in the flood plain. The diffusive wave approximation is used in both cases. In the flood plain, shallow wave theory is assumed to apply. Equations are solved by means of an Integrated Finite Difference approach on the basis of polygonal cells known as Dirichlet tessellation. The latter is derived from a suitable Digital Terrain Model (DTM) constructed with a particular Triangular Irregular Network (TIN) accounting for structure and break lines. Flood mapping and damage assessment procedures for agricultural land, traffic conditions and built-up areas have been implemented in a GIS framework supported by the data base system, dBaseIV, the raster GIS, IDRISI and the vector oriented support, MapInfo. Migration to more sophisticated GIS platforms is envisaged in the framework of future applications.

INTRODUCTION

This paper describes potential applications of GIS technology for flood mapping and damage assessment. This research is part of the EC-EPOCH-AFORISM project (A comprehensive Flood fOrecasting system for RISk Mitigation and control of the European community Program On climate and natural Hazards). The main objective of AFORISM is to develop an overall methodology for real time flood control. Rainfall forecasts are determined on the basis of radar, meteorological models and stochastic approaches. The expected precipitation is then transformed into flows with deterministic hydrologic models. Floods are then routed through the river network with hydraulic models to delineate flooded areas. Flood maps are combined with land use information to derive potential damages. These evaluations provide the basis for decision making with expert or heuristic systems. The comparison of various control alternatives should define the best management strategy for a given flood episode. This paper will only focus on the flow routing and impact assessment components of AFORISM. Although AFORISM is oriented towards real time control applications, the procedures presented herein can also be applied to land use planning.

Flood maps must describe the spatial variability of the hydraulic parameter(s) required by the subsequent damage evaluation procedures. For instance, in built-up areas the maximum water depth is the determinant parameter while agricultural productivity may be more sensitive to the total duration of submersions. Flood impacts on road networks must be evaluated according to the location of road cut-offs and the duration of traffic inconvenience. Overlays between flood and land use maps generate flood effects. Such maps show either the flooded buildings or the location of road sections where traffic circulation is no longer possible. Combining flood effects with worth criteria leads to a flood impact map and a numerical evaluation of flood damages. The latter can be estimated on the basis of an economic figure or according to indices. For built-up areas, it is relatively easy to compute economic damages according to land use maps and insurance statistics. Agricultural damages may be evaluated on a similar basis. Flood impacts on traffic require the development of an adequate index accounting for the average vehicle flow, the duration of flooding and the time delay between departure and destination points. Any change in the control strategy will lead to a new flood impact map and another flood damage evaluation. A multi-criteria analysis should finally determine the best control policy.

HYDRAULIC MODEL AND FLOOD DELINEATION

The proposed hydraulic model is based on a two dimensional diffusive approximation of the Saint-Venant equations:

$$\frac{\partial uh}{\partial x} + \frac{\partial vh}{\partial y} + \frac{\partial H}{\partial t} - q = 0 \tag{1}$$

For an infinitesimal element, equation (1) can be written in terms of flows:

$$\frac{\partial Q_x}{\partial x} dx + \frac{\partial Q_y}{\partial y} dy + \frac{\partial H}{\partial t} dydx - qdxdy = 0 \tag{2}$$

where:
H = water surface elevation at the control element (m)
u,v = velocities in the x and y directions (m/s)
h = water depth in the control element (m)
q = external inflows (m/s)
Q_x = flow rate in the x direction (m³/s)

The motion equations can be written as follows:

$$Sf_x + \frac{\partial H}{\partial x} = 0 \tag{3}$$

$$Sf_y + \frac{\partial H}{\partial y} = 0 \tag{4}$$

where: Sf_x = friction slope in the x direction

According to Manning's formula and equations (3 and 4), the following expression may be derived if the velocity components of u along y and v along x are neglected:

$$Q_x = -\frac{R_x^{2/3}}{n_x} hdy\sqrt{\frac{\partial H}{\partial x}} \quad Q_y = -\frac{R_y^{2/3}}{n_y} hdx\sqrt{\frac{\partial H}{\partial y}} \tag{5}$$

where: $R_{x,y}$ = hydraulic radius in the x and y directions
$n_{x,y}$ = Manning's roughness coefficient in the x and y directions
$Q_{x,y}$ = flow rates in the x and y directions

Substituting (5) into (2) leads to a partial differential equation which is only a function of H.

$$\left(\frac{\partial}{\partial x} K_x \frac{\partial H}{\partial x}\right) dxdy + \left(\frac{\partial}{\partial y} Ky \frac{\partial H}{\partial y}\right) dxdy + qdxdy = \frac{\partial H}{\partial t} dxdy \tag{6}$$

where:

$$K_x = \frac{R_x^{2/3} h}{n_x \sqrt{\frac{\partial H}{\partial x}}} \quad K_y = \frac{R_y^{2/3} h}{n_y \sqrt{\frac{\partial H}{\partial y}}} \tag{7}$$

Integrating equation (6) over a given domain S, leads to the following equation:

$$\iint_S \left(\frac{\partial}{\partial x} K_x \frac{\partial H}{\partial x} + \frac{\partial}{\partial y} K_y \frac{\partial H}{\partial y}\right) dxdy + \iint_S qdxdy = \iint_S \frac{\partial H}{\partial t} dxdy \tag{8}$$

Equation (8) can be integrated over a sub domain S_i of contour length L_i assuming that q and H are average values over S_i. The first term on the left hand of (8) can be transformed into a line integral resulting from an integration by parts and the application of the Green's theorem to eliminate the second derivatives of H.

$$\oint_{L_i} \left(K_x \frac{\partial H}{\partial x} \vec{x} + K_y \frac{\partial H}{\partial y} \vec{y} \right) . dL + q_i S_i = S_i \frac{\partial H}{\partial t} \qquad (9)$$

The term in the line integral describes the components in the xy co-ordinate system of the flow vector which is perpendicular to the contour limits (L_i) of the sub domain S_i. The line integral may be interpreted as the sum of all flow exchanges along the contour of the sub domain.

Consequently, equation (9) is a mass balance equation for the sub domain S_i since $q_i S_i$ is equal to the net inflow and the right hand side term represents the corresponding change in storage. If the entire domain is subdivided into polygonal cells using the Thiessen method and assuming that the mean properties of each mesh can be represented by its nodal point (i), it can be easily demonstrated that equation (9) can be written as follows:

$$\sum_{j \in M_i} K_{ij} B_{ij} \frac{H_j - H_i}{d_{ij}} + q_i S_i = S_i \frac{\partial H_i}{\partial t} \qquad (10)$$

where: M_i = set of nodes connected to node (i)
d_{ij} = distance between nodes (i) and (j)
S_i = polygon surface for node (i)
B_{ij} = interface length between nodes (i) and (j)
K_{ij} ⇒ is obtained from equation (7) when substituting the xy derivatives of H by the slope of the water surface profile along the [i↔j] direction which is perpendicular to the polygon facet, the length of which is B_{ij}.

Equation (10) can be solved with a time centred scheme where Δt is the selected time step, R_m and h_m are respectively the hydraulic radius and the water depth at the interface between nodes (i) and (j) and n_{ij} is the Manning's roughness coefficient along the (i↔j) direction:

$$\left[\left(\alpha \sum_{j \in M_i} \frac{R_m^{2/3} h_m B_{ij}}{n_{ij}} \sqrt{\frac{H_j - H_i}{d_{ij}}} \right) + \frac{S_i}{\Delta t} H_i \right]^{t+\Delta t} =$$

$$\left[\left(-(1-\alpha) \sum_{j \in M_i} \frac{R_m^{2/3} h_m B_{ij}}{n_{ij}} \sqrt{\frac{H_j - H_i}{d_{ij}}} \right) + \frac{S_i}{\Delta t} H_i \right]^{t} + \qquad (11)$$

$$S_i \left[(1-\alpha) q_i^t + \alpha q_i^{t+\Delta t} \right]$$

R_m and h_m have to be defined at the interface between control volumes along the ($i \leftrightarrow j$) direction. Consequently, it is necessary to assume a particular water surface profile between two connected nodes. According to the derivation of (11), the most coherent choice is to assume a linear profile between H_i and H_j. The unknowns of the problem are H_i at time $t+\Delta t$. Exchanged flows for each polygon are computed half time step ahead. Rewritten in a more convenient matrix form, equation (11) is solved by an iterative procedure based on forecasted values of H_i at the beginning of each time step. Solutions for H_i at time $t+\Delta t$ in the left side term of (11) are computed with a modified conjugate gradient method proposed by Todini [2]. Convergence is reached when the maximum difference between computed flow values for two successive numerical iterations are lower than a given tolerance. Stability criteria were initially proposed and are still under investigation.

As already mentioned the hydraulic model should be able to simulate one dimensional flood propagation in the main channel and two dimensional routing in the flood plain. In the channel, control volumes are defined according to the location of cross sections. For each channel cell, the surface S_i is calculated as a function of cross section characteristics and the length of the control volume L_i (figure 1a). In equation (11) transmissivities between channel nodes are computed according to the following expression:

$$T_{ij} = \frac{1}{n_{ij}d_{ij}\sqrt{\dfrac{H_j - H_i}{d_{ij}}}} R_m^{2/3} A_m \qquad (12)$$

where A_m is the wetted surface at the interface cross section of the control volumes for the channel nodes (i) and (j). The flow (Q) between nodes (i) and (j) is equal to:

$$Q_{ij} = T_{ij}(H_j - H_i) \qquad (13)$$

The direction of flow is defined by the sign of the difference between H_j and H_i. Figure 1a shows an example of a river reach with a sill and illustrates the corresponding layout of control volumes. The advantage of this approach is that the river DTM relies exclusively on traditional cross sections and a standard definition of hydraulic structures for which adequate formulas are used instead of equations (13) and (12). These specific flow values are substituted in equation (11). Vitalini [3] implemented adequate expressions for equations (12) and (13) applicable to most commonly found hydraulic structures such as weirs, sills, bridges, sluice gates, etc.

For flood plain nodes, and according to shallow wave theory, the hydraulic radius at the interface R_m is replaced by the water depth h_m and the flow surface at the same interface becomes $h_m.B_{ij}$ in equation (12). Special hydraulic equations to compute transmissivities between flood plain nodes can also be specified. The topography in flat areas can be easily compared to a set of compartments delimited by roads and rivers (break lines and structure lines). In case of flooding, flow spilled over the river banks fills up these compartments. The flood progresses from one compartment to another as a function of the relative difference between water surface elevation and the altitude of the delimiting obstacles. Consequently, the DTM must be derived from a Triangular Irregular Network (TIN). This TIN must account for break lines and ensure a proper connectivity between channel and flood plain nodes. The facets of the triangles can intersect neither break lines nor structure lines. Each single triangle satisfies the condition that all internal angles are acute.

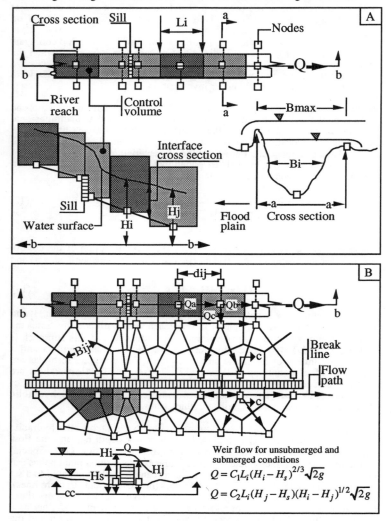

Figure 1.   Main characteristics of the hydraulic model

Figure 1b also shows the way in which river cells are connected to flood plain polygons. Lateral spilling over river banks is simulated with a broad crested weir equation where the length of overflow corresponds to the sum of the half way distances between node (i) and those located immediately upstream and downstream. Water depth over the bank is assumed to be constant. The weir loss coefficient is selected in order to limit the maximum overflow depth. The computation of flow between the channel and the flood plain accounts for both unsubmerged and submerged conditions. Figure 1b also illustrates that break lines are simulated in a very similar manner.

Vitalini [3] proposed a semi-automatic method to derive the Triangular Irregular Network (TIN) on which the polygonal mesh, required to solve equation (11), can be delineated. This method ensures that river and flood plain cells are properly connected and provides an immediate estimation of the overflow length in the weir equation in case of bank overtopping.

This way of handling lateral overflows from rivers as well as break lines simplifies the construction of the DTM since these features do not have to be explicitly defined. They are implicitly accounted for in the hydraulic calculations by selecting an adequate equation to compute transmissivities. The necessary parameters may be derived directly from the topographic measurements. The latter are taken in a particular manner according to the requirements of the proposed hydraulic model.

Figure 2a shows an example of a TIN construction for a 3 km^2 area in the Basse Broye river flood plain region located in the Swiss plateau region. Figure 2b illustrates the corresponding polygonal mesh and the main parameters used in equation (11).

In this example, river flow is considered to be predominantly one dimensional and exchanges with flood plain polygons are only possible through bank overtopping. It can be noticed that no control volumes are needed to define break lines. Their influence on flow exchanges are implicitly accounted for in the solution of equation (11). At each time step, the model computes water surface elevations at the nodal point of each polygon (figure 2c). This information can be rasterised with an adequate routine to produce a flood map similar to that shown in figure 2d.

FLOOD DAMAGES TO AGRICULTURE

Flood damages to agriculture result from more or less prolonged submersions affecting the normal growth rate of plants. According to the period of flooding and the duration of submersion, the farmer may decide to accept either a partial, even a complete, damage or seed a replacement crop if the season and/or the soil conditions allow for it. The duration of submersion is defined as the time span between the beginning of flooding and the drying out of the soil. Until then, the soil and the plants may be damaged by labour or heavy machinery.

One of the most important aspects was the handling of the time component since the location of crops varies from one year to another according to predefined rotation rules and the vulnerability of plants changes within the growing cycle between seeding and harvest periods.

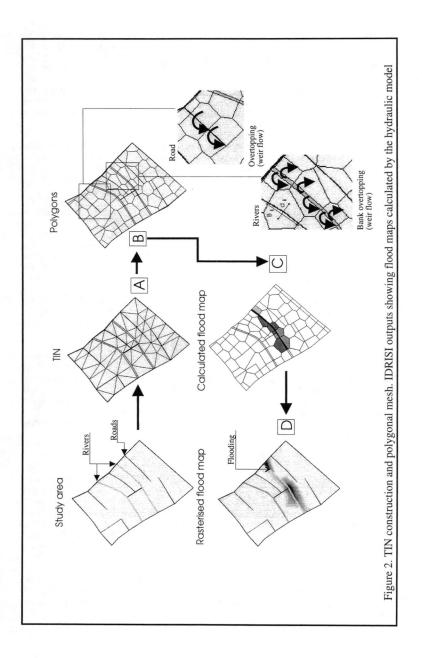

Figure 2. TIN construction and polygonal mesh. IDRISI outputs showing flood maps calculated by the hydraulic model

Figure 3 illustrates the overall set up in MapInfo and dBaseIV to evaluate flood damages to agriculture. Figure 6a shows the study area with the plots and the crop types implemented in the field when flooding occurs. Plots appearing in grey have crops which are not sensitive to submersion at that particular period. The total durations of submersion for flooded polygons as computed by the hydraulic model are also shown in figure 3a. Figure 3b illustrates the logical sequence of operations in the database system. Polygon # 6 corresponds to plot # 42. The plot surface flooded is equal to 0.1925 ha and the duration of submersion is equal to 8 days.

The "Flooded Plot" table was generated by overlaying the map showing total durations of submersion and that describing plots and crop types at the moment of flooding. For this particular event, plot # 42 was occupied by crop # 7 (Crop-Plot table) which corresponds to seed pasture with a profit margin of 2990.- SFr/ha (Crop Information table). On the basis of the duration of submersion (8 days) and the crop number (# 7), the system identifies potential losses with and without replacement in the "Potential Loss" table. If the new crop identifier is equal to zero, no replacement is possible and damages in both cases are obviously identical.

If the duration of submersion is equal to zero, the actual crop has no damages and replacement is of course not required. For crop # 7, potential losses appear to be higher for the replacement option. To compute effective losses, the system compares both options on the basis of the flooded surface for the no replacement option and according to the total plot surface in case of replacement. For the example shown in figure 6b, the cheapest option is to accept actual damages which amount to 0.1925 ha*2025.- SFr/ha $\cong$ 390 SFr. This number is then stored in the "Crop-Plot" table on which the flood damage map in figure 3c is based.

FLOOD IMPACTS TO BUILT-UP AREAS

For a given flood episode, the proposed hydraulic model computes water depths at each time step and for each polygon. A pre-processing routine in dBaseIV identifies for each node the maximum water depth during the entire flood event. This information is transferred to IDRISI to produce a map showing the spatial distribution of maximum water surface elevations. Overlaying this map with that containing the location of individual buildings identifies all the constructions affected by flooding.

IDRISI computes the average depth over the surface covered by each single building. If this average depth is higher than a critical value, the building is completely destroyed. Losses are equal to the value of the structure and that of the building content.

If the water depth is less than the critical value, losses are estimated on the basis of stage-damage relationships. In that case, damages are proportional to the value of the building content. All these computations are performed within dBaseIV. Damages for each single building are transferred to IDRISI to produce a map showing the spatial distribution of economic losses. Finally, dBaseIV cumulates individual damages to produce the global flood impact on built-up areas.

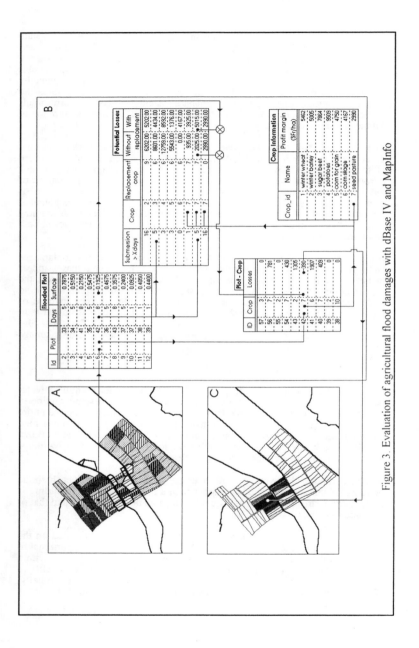

Figure 3. Evaluation of agricultural flood damages with dBase IV and MapInfo

Typical output maps produced by IDRISI are shown in figure 4. Figure 4a shows the spatial variability of maximum depths for each polygon and indicates all the buildings affected by flooding. Figure 4b illustrates the resulting damages expressed as a function of the total value of each single concerned building.

FLOOD IMPACTS ON TRAFFIC

Traffic damages relate to flow interruptions and to reductions of circulation speed. In case of road cut-off, the search for the fastest alternative path between two points is a crucial question for rescue services. The effects of traffic interruptions can be evaluated on the basis of minimum travel times between departure and arrival points in the road network. The traffic network can be subdivided into various categories according to average speed and traffic conditions (national, cantonal, local, agricultural tracks, etc.). Each road is made up of a series of sections. A section is defined as a piece of road with no intersections. A section is bounded by two cross-roads.

According to road network categories, each section is characterised by a given travel time. Between two points in the network, it is possible to compute the shortest travel time (Dopt). The corresponding path is defined as a principal axis. It always coincides with a major road between two cities or important sites. These axes, including the associated sections, are stored in the database system as non geographic elements. They are also characterised by an average daily traffic flow derived from regional development plans commonly available in Switzerland. The proposed methodology assumes that travel times are equal in both directions of circulation. The average flow per axis is equal to the sum of those in each traffic direction.

In case of flooding, the system identifies the road sections where circulation is no longer possible. Traffic cut-off occurs if the depth of submersion on the road is higher than a given value and lasts longer than a pre-specified interval.

For selected departure and destination points, the system re-computes new travel times and identifies the fastest paths. For each axis the impact of flooding on traffic flow can be computed with the following formula:

$$N = \left(1 - \frac{Dopt}{Dino}\right)FD \qquad (14)$$

where Dino is the travel time in case of flooding, F is the average vehicle flow in normal conditions and D the duration of flooding. N is computed for each axis. It represents the number of vehicles that could not reach destination during the flooding period.

If N=0, the axis is not concerned with flooding. The maximum value of N is equal to FD which means that the axis is cut off for a long lasting period.

The overall impact of flooding is derived by cumulating the individual values of N for each axis. The system will also compute the new spatial distribution of traffic density resulting from flooding. This allows to identify overloaded sections and to suggest alternative paths.

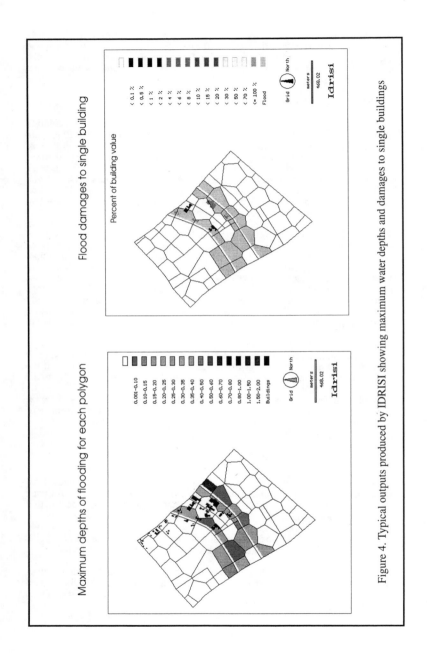

Figure 4. Typical outputs produced by IDRISI showing maximum water depths and damages to single buildings

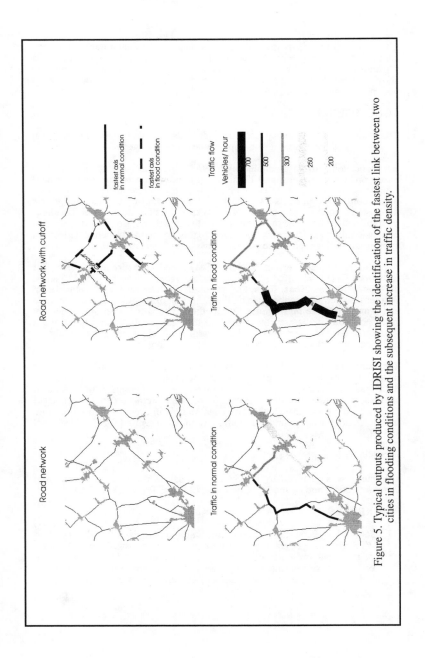

Figure 5. Typical outputs produced by IDRISI showing the identification of the fastest link between two cities in flooding conditions and the subsequent increase in traffic density.

The new spatial distribution of traffic is computed by cumulating for each single section the vehicle flows of all axes passing through it. Figure 5a illustrates the search for an optimal path in case of flooding cut off. Figure 5b shows a typical output produced by IDRISI comparing traffic densities for both normal and flooding conditions.

CONCLUSIONS

The study presented in this paper clearly demonstrates that GIS technology can be applied to flood mapping and impact assessment problems.

For flood delineation, a suitable hydraulic model has been developed. It is based on a two dimensional diffusive approximation of the Saint-Venant equations. The latter are solved with a time centred finite difference approach using control volumes. In the channels, flood routing is essentially one dimensional and relies on traditional cross sections and a standard definition of hydraulic structures. In the flood plain, the equations are solved on a two dimensional basis. Channels and flood plains are properly connected by an appropriate delineation of the polygonal meshes (Thiessen type).

A particular Digital Terrain Model (DTM) based on a Triangular Irregular Network (TIN) is required to produce the polygonal meshes needed to solve the hydraulic equations. This TIN accounts for structure and break lines. Further research will concentrate on the definition of improved stability criteria and a more reliable mesh generation method. The latter is an open question since polygonal cells are constructed on the basis of distance criteria and a triangular irregular network. Other topographic characteristics should be included such as local slopes or preferential flow paths.

The GIS framework includes the database system dBaseIV, IDRISI for the raster component. and MapInfo for the vector support The GIS prototypes described in this paper were initially designed as demonstration tools to analyse feasibility problems and data accuracy requirements. Joerin [1] developed the necessary Conceptual Data Models in order to migrate to more sophisticated platforms under a UNIX environment using ORACLE, SPANS and VISION or ARCINFO.

ACKNOWLEDGEMENTS

This study was financed by the Swiss Foundation for Scientific Research; Grant # 510047; and by the EC, Contract # EPOC-CT90-0023 (TSTS).

REFERENCES

[1]   Joerin, F. (1993). Conceptualisation statique et dynamique d'un système d'information géographique utilisé comme outil de simulation. Mémoire de recherche présenté pour l'obtention d'un certificat de cours postgrade en bases de données. Département d'Informatique de l'Ecole Polytechnique Fédérale de Lausanne; Lausanne, Suisse, 80 pp.

[2]   Todini, E. (1991). Hydraulic and Hydrologic Flood Routing Schemes. Recent Advances in the Modelling of Hydrologic Systems; Kluwer Academic Publishers.

[3]   Vitalini, F. (1993). Una procedura per la previsione, il controllo e la gestione del rischio di inondazione in forma distribuita sul territorio. Tesi di Laurea. Facolta di Ingegneria, Dipartimento di Ingegneria Idraulica, Ambientale e del Rilevamento, Politecnico di Milano, Milano, Italia, 300 pp.

Object-Oriented Finite Volume Dam-Break Model

Bochra Ech-Cherif El Kettani[1] and Driss Ouazar[2]

(1) Centre National de Coordination et de Planification de la Recherche
   Scientifique et Technique (CNCPRST) - BP 8027, Agdal, Rabat, MAROC
(2) LASH, Laboratoire associé au CNCPRST, Ecole Mohammadia
   d'Ingénieurs, B.P 765 - Agdal, Rabat, MAROC.

ABSTRACT

This paper is devoted to Object-Oriented analysis and design techniques to two computational modules in the field of dam-breaking.
The C_{++} Object-Oriented language is used. After the problem partionning into useful and adequate classes, the codes are developed, benefiting from the main features offered by Object-Oriented Programing, namely : modularity, data abstraction, inheritance, polymorphism and automatic memory allocation. When well-managed, object-oriented programming is found to be very promising as for programmer productivity, software flexibility and reusability, and software maintability.

INTRODUCTION

Few 2D commercially available software for dam-break analysis exist. This shortness is mainly due to :
. high cost of software development and maintenance even if highly sophisticated numerical techniques are well-established including Finite Difference Methods, Finite Element Methods and Finite Volume Methods ;
. the software should embody a good data management, knowledge representation and graphic capabilities and/or Geographic Information System coupling ;
. past hardware limitations ;
. multiplicity of software environments with various possibilities and limitations.
This work is a part of an effort to develop a software methodology for dam-break wave assessment and analysis.

The paper is devoted to the introduction of object-oriented analysis and design techniques to an existing Finite Volume Dam-Break model [1], the overall objective being the development of an Object-Oriented Expert Alarm System in the dam-breaking field, with involvement of pre and post-processing facilities.
Firstly, we present the overall architecture of the software package describing each of its components, then, a presentation of the work under development is made.

PROJET DESCRIPTION

Objectives : The purpose of the present project is to establish reliable maps of inundated and dry areas at the downstream of a collapsing dam. Such maps allow to prepare civil protection plans which are of paramount importance when dam failure danger cannot be avoided by taking some suitable measures, like a rapid reservoir emptying. In this case, a preventive evacuation of populations threatened by an imminent dam collapse is the only effective mean, provided that it is done in time. This requires the establishment of an appropriate alarm plan based on reliable estimations of both downstream flow depth and devastating flood wave arrival times.

Methodoly : Dam-break wave analysis involves two kinds of knowledge :
- an algorithmic knowledge, associated with the computation of physical processes governing the dam failure process. This kind of information consists in methods and procedures evaluating the unknown problem quantities (water depth, water velocity, discharge through the breach or over spillways, etc...).
- an heuristic knowledge, consisting of a set of fragmented and scattered information, rules of thumb which often have their origins in experiments and skilled advice of field experts.

Field experts'know-how being difficult to model by means of definite and sure algorithms, the methodology of Knowledge Based Expert Systems - one of the most important Artificial Intelligence application - is found to be particularly suitable to reproduce human judgement and decision faculties, and to deal with some intellectual tasks like catastrophs prediction or the conception and planification of a prompt alarm action.
Such methodology will account elegantly for the hybrid aspect of knowledge involved in dam-break wave assessment and analysis. We describe in the following section the Expert Alarm System, provided with user friendly data input and output tools.

SOFTWARE ARCHITECTURE

The expert system will consist of an inference engine operating on a knowledge base and computational modules, with user friendly pre and post - processing tools.
The different components of the software package are reported on fig. 1. Details are shown on fig. 2.

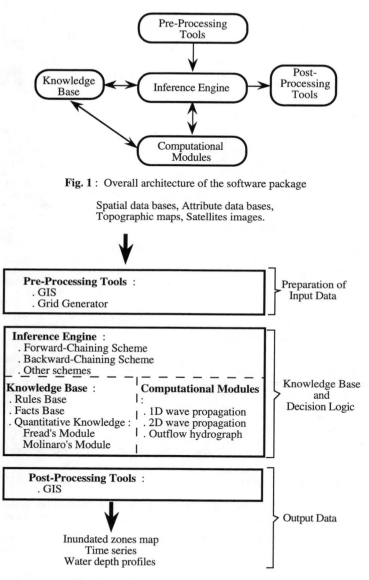

Fig. 1 : Overall architecture of the software package

Fig. 2 : Software package architecture details.

Pre-Processing Tools

Data Digitizing Modules : Topographic data and soil features are of paramount importance in dam-break wave propagation computations. Topographic data include data available on spatial databases (spatial localizations of soil features), attribute databases (soil features : slope, roughness...), topographic maps, satellites images and/or photographies.

This geographic information needs in a first step to be digitized and/or scanned, and in a second step efficiently managed and analyzed.

Acquisition, storage, analysis and representation of topographic data will be dealt with within a Geographic Information System (GIS), thus allowing maps digitizing, geographic databases management, geographic analysis and cartographic representation on various output devices (screen, printer, plotter).

Grid Generation Module : Quadrilateral Finite Volumes are used for the representation of the flow field. The quadrilateral mesh generator will be operating on the digitized database.

After the flow field discretization into quadrilateral elements, the corresponding spatial and attribute input databases will be prepared. The set of x-y-z cartesian coordinates associated with grid nodes will be stored by means of the GIS, as well as the soil features values which will be deduced at each mesh node by interpolating the digitized database values.

The Knowledge Based Expert System

The knowledge BasedExpert System is a relatively general Inference Engine handling both algorithmic and heuristic knowledge stored in the Knowledge Base.

The Inference Engine : Given the initial data and the problem parameters values, the Inference Engine will lead an heuristic search, using the information available in the knowledge base and computational models. In other words, the Inference Engine will be choosing the knowledge to be applied, the search progressing in a non-exhautive way.

The Knowledge Base : When the Inference Engine starts to operate, the Knowledge Base embodies :
- facts to be established (goals) and recognized facts : this is the facts base ;
- heuristic knowledge representing experts know-how and often expressed in terms of rules : this is the rules base.

Rules will be expressed in terms of "If... then" statements.

Reference to computational modules will appear as function calls in the right (or left) side of the "If ... then" statement.

When operating, the Inference Engine will continually process information in the knowledge base (addition, retrieval and withdrawal of information).

The Computational Modules : Dam-break wave analysis involves two important aspects (from the algorithmic point of view) :
- the generation of the outflow hydrograph, given the dike geotechnical features and breaching process ;
- the study of the wave propagation at the downstream of the dam.

Dam-break wave computation :
Governing flow equations are the 2D shallow water equations, accounting for 3D effects by means of depth averaging process. They are solved using the Finite Volume Method, based on the Mac Cormack Scheme, a second order explicit scheme operating on a trapezoïdal finite volume. The stability of the numerical scheme is ensured by applying the CFL condition.

Outflow hydrograph calculation :
The outflow hydrograph is either given as input by the user or generated by means of the following module at the dam site, given the dam breaching process. This module uses a hydrologic routing technique,based on the principle of mass conservation and accounting for the reservoir storage characteristics, the reservoir inflow and total outflow. The total outflow discharge involves the outflow due to erosion phenomena (breaching, piping) and the outflow over designed outlets (spillways, dam crest).
The breach model is taken herein from Fread [3]. Other models exist (for example, Giuseppetti and Molinaro's model [4] ...), and can also be used to compute the outflow hydrograph.

Post-Processing Tools :

The results will be displayed on the topographic maps as :
   - velocity vectors to indicate flow direction and magnitude at each grid node, for a given time step.
   - contour lines to delimit inundated and dry areas, for different time steps. Each contour line will be labeled with the corresponding wave arrival time, thus allowing to follow in space and time the devastating wave evaluation.
To process and visualise obtained results, the GIS could be used, as well as some graphical software like GKS for time series (depth and velocity) at a given location, and/or water depth profiles along the channel.

PRESENT WORK : INTRODUCING THE OBJECT-ORIENTED PARADIGM

After this description of the whole Expert Alarm System as it has been designed, we will consider more closely the computational modules simulating respectively dowstream wave propagation and outflow hydrograph. The first module, described in a previous work [1,2], was coded in a procedural programming style, using Pascal language. We are at present undertaking a complete restructuring of each code, using C_{++} language [5,6] and Object-Oriented concepts [7,8]. The C_{++} language is the Object-Oriented extension to C developed at AT & T by Bjarne Stroustrup. It is a hybrid language wedding object-oriented functionality with the features of traditional and efficient structured languages. C_{++} will provide the programmer with object-oriented capabilities without loss of memory efficiency or run-time.

The Object-Oriented Paradigm :

An Object-Oriented program is partitioned into modules or components containing both data and procedures : the Objects.

Object-Oriented Programming (OOP) is thus an approach centered on objects, while traditional programming is procedures-oriented. In other words, when programming in an object-oriented style, the main program unit is managing and controlling a set of objects, not procedures.

After introducing this basic concept of object-based modularity, the main features provided by Object-Oriented programming will be reviewed through our two applications, after partitionning each problem into several objects types called classes. The benefits of moving from procedural to Object-Oriented programming are discussed.

An Object-Oriented Module for Dam-Break Wave Propagation

When simulating dam-break wave propagation using the Finite Volume Method based on the Mac Cormack scheme, one needs to :

- consider the computational field with all its features : Valleys class is thus defined
- calculate water depth and velocity at each grid node, handling a set of matrixes which represent space, topography and flow data : Matrixes class is therefore created, containing all useful matrix functionalities. Fig. 3 shows the corresponding simulation classes.

This first application illustrates quite well what classes do represent in general :

. concepts in the application domain, like the class Valleys ;
. pure artifacts of the implementation, like the Matrixes structure.

Matrixes Class	Valleys Class

Member Data :
. matrix dimensions
. matrix elements
 	
Member Functions :
. Constructor·data initialisation
  and matrix-allocation
. Destructor·matrix de-allocation
. matrix element storage
. operations on matrixes
  (+, x, inverse, ...)
	Member Data :
	. mesh nodes matrix
	. topographic data matrixes
	. flow variables matrixes
	
	Member Functions :
	. data initialisation
	. flow variables prediction
	. flow variables correction
	. boundary conditions

Fig. 3 : The simulation classes for the downstream wave propagation.

Information hiding and Data abstraction :

Given the class definition as a set of member data and member functions, each class instantiation (i.e. each associated object) contains hidden inside its data structures and methods characterizing its behaviour.

Consider for instance the class Matrixes : for a specified matrix, the object state information is stored in matrixes class in the form of internal data ; the access to the matrix state variables is allowed only for the member functions of Matrixes class, implementation details being hidden from the class Valleys. This property is known as information hiding.

In turn, an object valley (instantiation of the class Valleys) cannot access directly a matrix data structure. The access can be forced only by invoking one of the member functions of the class Matrixes.

This property is known as data abstraction.

Information hiding coupled with data abstraction will allow to change the internal representation of a class, without having to change the other classes implementation, thus leading to a better system reliability and a better organisation of the inherent problem complexity.

Automatic Storage Management :

C_{++} language allows automatic memory storage for dynamic objects by means of the following two operators : "new" and "delete". Consider for instance the class Matrixes : the "new" operator is used in the member function Constructor to allocate automatically memory for a two-dimensional array. We can also free up automatically the memory allocated for the matrix object, using the "delete" operator in the member function Destructor : the cleanup work is thus performed after an object (the matrix) is no longer needed. Automatic storage management will increase software reliability.

An Object-Oriented Module for Outflow Hydrograph Computation

Let us consider the second computational module that estimates the outflow discharge at the dam site. As shown in fig. 4, we have defined for this system two classes : the class Reservoirs and the class Water_Outlets.

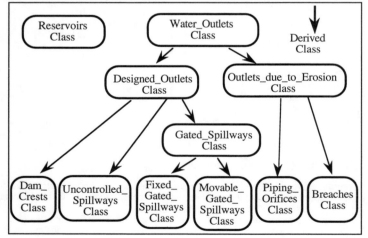

Fig. 4 : The relationship between the simulation classes of
the outflow hydrograph

Inheritance :

As we need to instantiate objects for water outlets, each with its own private data, we define separate types of water outlets as different classes : they can therefore inherit common properties and still possess their own special features. This property of inheritance is for instance illustrated in detail on fig. 5 where the class Outlets_due_to_Erosion and its two derived classes Piping_Orifices and Breaches have been reported : indeed, an orifice equation is different from the equation governing flow hydrodynamics over a breach (the breach being approximated by a broad-crested weir). But both the orifice and the breach are due to progressive erosion phenomena, reach a final width b and a final bottom elevation Hbm after a development time Tfb til complete failure.
Inheritance allows to reuse classes by sharing code between one class and its subclasses, thus reducing considerably program size and development time.

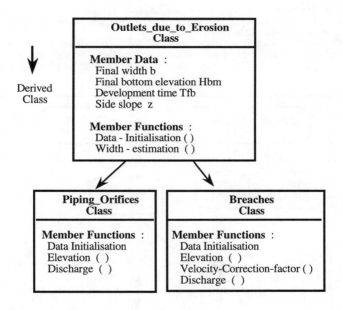

Fig. 5 : Detailed relationship between the Outlets_due_to_Erosion class
and its derived classes

Polymorphism :

Considering once more the same system for outflow hydrograph generation, we can notice that the classes Piping_Orifices and Breaches implement a method under the same and unique name : discharge(), but with two different implementations (because of the different hydrodynamic equations in play). This illustrates the ability of the member function call x.discharge(), to invoke different procedures, depending on the specified class for object x ; in other words, the version of discharge() that will be invoked is that one corresponding to the type of object x.
This property, known as polymorphism, allows to addition new types of objects without modifying existing procedures, thus increasing the software flexibility.

Conclusion :

From this application, we can conclude that both inheritance and polymorphism will allow an easy extensibility of the code, and will simplify the maintenance of large object-oriented modules.
We can illustrate this by trying to extend the code realized for the previous application (dam-break wave computation), and this by adding the module solving the 1D case. Fig. 6 shows the corresponding set of simulation classes.

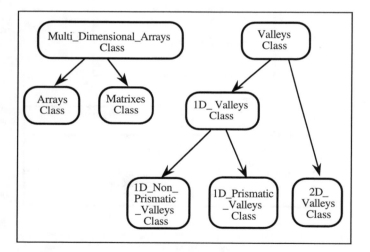

Fig. 6 : Extension of the code computing downstream wave propagation

Object-Oriented Programming and Artificial Intelligence

Object-oriented programming provides a powerful possibility of knowledge representation.
Indeed, we can use Object-Oriented concepts to organize the knowledge base into sets of specific rules. This would facilitate the heuristic search strategy lead by the inference engine, since the search will be limited to a set of rules amongst others [9,10].
A logical continuation of this work would be the realisation of an object-oriented knowledge base.

CONCLUSION

We propose herein a restructuring of a Finite Volume Dam-break code, using C_{++} language and object-oriented concepts. After the problem partionning into useful and adequate classes and objects, the corresponding code is developed, benefiting from the main feactures provided by OOP :
   . an object-oriented modular structure, partitioning the code into classes.
   . a support of data abstraction types for information hiding.
   . the ability to share code through inheritance.
   . a dynamic polymorphism that enables addition of new types of objects
      without modifying existing procedures.
   . an automatic storage management.

When well managed, Object-Oriented design is found to be a better alternative for software development since abstraction and information hiding result in a better organisation of the inherent problem complexity, class reuse enables a simplified system integration, and inheritance and polymorphism allow easy extensibility of codes.
The use of object-oriented paradigm seems very promising as for programmer productivity, software flexibility and reusability, and software maintability. Futhermore, OOP provides a powerful possibility of knowledge representation leading to great efficiency when coupling numerical techniques and artificial intelligence concepts.

REFERENCES

[1]   Ech-Cherif El Kettani B., Berrada A., Ouazar D. and Agouzoul M. (1991). "2D Dam-Break Flood Wave Propagation on Dry Beds", Proc. 2nd Int. Conf. on CMWR, Oct. 7-11, Rabat, Morocco..

[2]   Ech-Cherif El Kettani B. and Ouazar D. (1993). "2D Dam-Break Flood Wave Propagation", Proc. 1st Int. Conf. on Hydro-Science and - Engineering (1st ICHE), published under the title "Advances in Hydro-Science and -Engineering", ed. Sam S.Y. Wang, Vol. I, pp. 775-782, June 7-11, Washington D.C.

[3] Fread D.L. (1982) "DAMBRK : The NWS Dam-Break Flood Forecasting Model", National Weather Service, Office of Hydrology, Silver Spring,Md.

[4] Giuseppetti G. and Molinaro P. (1989). "A Mathematical Model for the Erosion of an Embankment Dam by Overtopping", Int. Symp. on Analytical Evaluation of Dam Related Safety Problems, Copenhagen.

[5] Stroustrup B. (1991). "The C++ programming Language : Second Edition", Addison-Wesley Publishing Compagny.

[6] Weiskamp K. and Flamig B. (1989). "The Complete C++ Primer", Academic Press Inc.

[7] Wiener R.S. and Pinson L.J. (1988) "An Introduction to Object-Oriented Programming and C++ , " Addison-Wesley Publishing Company.

[8] Masini G., Napoli A., Colnet D., Leonard D. and Tombre K. (1990) ."Les Langages à Objets", InterEditions, Paris.

[9] Farreny H. and Ghallab M. (1987). "Eléments d'Intelligence Artificielle", Editions Hermès, Paris.

[10] Tello E.R. (1989). "Object-Oriented Programming for Artificial Intelligence : A Guide to Tools and System Design", Addison-Wesley Publishing Company.

A Stochastic Framework for the Modeling of Failures in Urban Drainage Systems due to Microscale Effects

P. La Barbera, L. Lanza and U. Parodi
University of Genova, Institute of Hydraulics
Montallegro, 1
16145 Genova, ITALY

ABSTRACT

A stochastic approach to the modeling of urban drainage processes at the scale of landscape perturbations is proposed in this paper aimed at evaluating the actual drainage efficiency in preventing local flooding due to microscale effects. Local failures are indeed observed as a consequence of precipitation events - even those not exceeding the design rainfall - when gully holes at random positions are forced to drain contributing areas which are larger than the design ones by the fact that the system of connections between the surface and subsurface drainage network fails and water is driven away from design flow patterns. The assessment of the role which is played by microscale perturbations in determining deviations of the runoff flow from the artificial drainage network is the major aim of the present work. A stochastic framework is proposed in this respect, where the process characteristics are defined on the basis of the assumed self-affinity features of urban landscape. Applications are expected in the identification of urban areas which are prone to flooding during heavy rainfall events as well as in the assessment of related damages and their spatial distribution.

INTRODUCTION

The basic concept of traditional urban hydrology relies on the coupling of a surface drainage network - artificially constrained by the layout of streets, pathways and further structural features of the urban landscape morphology - with a subsurface drainage system - made up of pipes and channels - specifically designed in order to lead precipitated waters quickly and effectively towards their outflow destination. The correspondence of the two systems is not expected to hold - apart of the general outline - and connections are ensured by a number of gully holes which are designed for draining well specified areas. The matching between simulated and measured hydrographs at the outlet of the main collectors is ususally assumed to be a suitable criterion for the assessment of the system efficiency.

During high intensity rainfall events, however, the system may prove locally inefficient - even if design rainfall is not exceeded - as a consequence of the unexpected amount of water which happens to reach a certain number of gully holes in the network. The latter are indeed forced to drain a contributing area which is larger than the design one by the fact that the system of connections between the surface and the subsurface network fails and water is driven away from design drainage patterns. Microscale effects due to the landscape anthropization are usually responsible for local failures of the urban drainage system.

Both the lack of knowledge about the detailed morphological configuration of the urban landscape and the computational effort which would be required for the management of the eventually collected data makes the modeling of urban landscape features at the resolution of such microscale perturbations obviously unsuitable within a conventional approach to the problem. Nevertheless, the high resolution texture of the urban morphology might be reproduced within a stochastic approach, provided that the latter is supported by a given underlying skeleton which is deterministically observed at a larger scale.

The potential of simulating the effect of microscale components on the basis of a suitable description of the landscape and a set of stochastic modeling procedures is investigated within the paper. In Section 2 the formulation of the problem is proposed in the framework of a brief review of the assumptions actually underlying the traditional design criteria. The problem of resolution which limits the actual knowledge of the urban landscape is discussed in Section 3. In Section 4 the stochastic representation of microscale failures is described while the simulation procedures are proposed in Section 5. Conclusions are drawn with reference to the practical applications which are expected in the field of urban drainage modeling as well as for the prevention and management of local floodings in highly urbanized areas.

THE TRADITIONAL APPROACH

Storm sewer design and verification traditionally require three major modeling steps, namely the evaluation of rainfall inputs over the urban catchment of interest, a model for transferring precipitated water to the gully holes, and a model for the flow routing in the underground network. The precipitation process presents a strong variability both in space and time. At the scale of urban catchments (a few square kilometers) it is difficult to assess the spatial variability of precipitation on the basis of the available monitoring devices. The rainfall is therefore usually considered uniformly distributed in space over the catchment domain and the stochastic charac-

teristics of rainfall are taken into account only in regards to its temporal distribution. Two approaches are generally used:
- simulation using historical records;
- creation of design storms associated to their return periods.

Most of the methods based on design storms use, as a starting point, the IDF (Intensity - Duration - Frequency) curves obtained from frequency analysis of rainfall data measured in the study area or by means of empirical formulas obtained for neighboring or similar catchments (Chow [1]) From IDF curves design storms are obtained using various techniques. For all of them the typical storm duration has to be chosen by the designer in relation to the concentration time of the basin of interest or by analyzing storm series. The storm shape can be predefined as in the case of the triangular hyetograph method or directly established from the IDF curves as for alternate black box methods. Nevertheless the above mentioned methods don't reproduce the storm shape corresponding to a given frequency of occurrence, although recognizing that the shape of the storm is quite important in determining the response of the basin to the rainfall solicitation. A design hyetograph obtained from storm analysis (Huff [2]) can relate intensity distribution patterns to the frequency of storm occurrences.

The areas drained by a given gully hole are determined by analyzing flow paths on the basis of the knowledge of natural drainage and roadwise patterns. The method is deterministic and the influence areas time invariant. In general, due to the difficulties of describing the landscape morphology at a suitable resolution scale in complex urban contexts, the traditional approach can give rise to doubtful drainage area attributions. Infiltration is usually determined by the runoff coefficient method (Pilgrim [3], Linsley [4]) even if more complex techniques - such as the Horton's model - are sometimes applied to take into account the variability of the basin response due to time changes in soil moisture content (Watt and Kidd [5], Voorhees and Wenzel [6]).

The surface routing of runoff waters is usually performed by means of lumped models based on different techniques (Sarma *et al.* [7]):
- the Rational Method;
- the Instantaneous Unit Hydrograph;
- the single or multiple reservoir methods;
- the combined linear channel and linear reservoir methods.

Distributed models for surface runoff - based on the data handling capabilities of Geographical Information Systems - have been applied on experimental basis for small urban watersheds (Smith [8], Ando *et al* [9]). The approach, based on the quasi-physically based modeling of runoff over inclined planes and along gutters are difficult to use in practice since the description of the topography is required at the scale of the micro structures. Most often the topographic information is available only at much larger scales (Djokic and Maidment [10]). In heavily urbanized areas the complexity of the street layout and of the micro structures makes the application of a distributed surface routing method impossible within an operational context.

The routing through the underground drainage network has been widely studied. The models applied in practice range from simple time shift schemes to models using the discretized full S. Venant equation. It is not the purpose of this paper to describe the various subsurface routing methods for this component of urban drainage modeling as the latter is not influenced by the use of a stochastic approach to urban hydrology. In summary, traditional storm sewer design and verification methods take into consideration the stochastic characteristics of rainfall input along the temporal domain approaching at the same time all further operational steps in a fully deterministic way.

THE DESCRIPTION OF URBAN LANDSCAPE

Due to the constraints imposed by computational tools and by the cost of data acquisition, the resolution scale of the landscape representation is selected as a compromise between the need for a suitable description of the investigated process and the data management requirements. In the case of urban hydrology the representation scale is necessarily coarser than the scale of variability of the morphological features of the landscape which actually affect the artificial drainage processes within the system. A few examples of the small scale description of urban catchments may be found in the literature (Hollis and Ovenden [11]): the approach is however limited to research purposes and not operationally applicable to large urban settlements.

A range of resolution scales is involved when dealing with the description of the urban landscape:

- the quarter scale;
- the block scale;
- the single house scale;
- the microscale.

An example of such a hierarchization of scales is provided in Figure 1 - from the quarter to the single house resolution - with reference to a portion of the urban settlement of the city of Genova (Italy).

Self-affinity characteristics may be recognized to hold when moving from one scale to another, obviously depending on town planning parameters as a function of the different types of urban structures. The modeling scale is the one which proves to be consistent with the landscape description capabilities, i.e. - generally - the block scale. The interior of each block is made up of morphological details (the single house scale) which are here assumed to reproduce the same texture of the large scale description and thus inferred from the known deterministic structure of the latter. In this view, the reconstructed description of the small scale morphology within each block is also assumed to reproduce - in a statistical sense - the same drainage properties as the large scale process.

At any finer resolution the description of microscale features (sidewalks, stairs, walls, gully holes unefficiencies, etc.) and that of the eventually damaged entities (shops, boxes, underground stores, etc.) relies on a stochastic representation which ensures - on average over a suitable number of realizations - the statistical consistency with the investigated process. The form and parameters of the probability distribution of such entities are to be derived from the observed self-affinity characteristics of urban landscape by a suitable correlation.

The method which has been briefly outlined in the present section allows the statistical reproduction of the deterministic structure of the landscape at any smaller scale than the observation one where the deterministic knowledge of the actual morphology is unrealistic due to the complex nature of urban landscape. Such a schematization is the basis for the analysis of drainage structures as addressed in the following sections.

THE DRAINAGE STRUCTURES

Three drainage networks will be considered in the present analysis with reference to the structure of urban systems, namely the natural, the artificial - or surface - and the subsurface network. The natural system is defined by the paths of steepest slope as determined automatically on the basis of the topographic information stored in a Digital Elevation Map (DEM) at a suitable resolution for the investigated area. Algorithms for the automated detection of the natural drainage morphology has been proposed in the literature by several authors (O'Callaghan and Mark [12], Band [13]). The procedure used in this work was presented by La Barbera et al. [14] in the case of catchment hydrology applications. In Figure 2a the outline of natural drainage patterns is superimposed to the urban landscape of the city of Genova for that portion of the territory already identified in the previous section. The Hortonian hierarchization of the network is also represented by means of different line thicknesses corresponding to the order of each link.

The artificial network is defined by the layout of streets and pathways which makes up the actual landscape all over the urban domain. This reflects the urbanization criteria used during the different stages of development of the investigated urban settlement. Only the large scale detection of such network is possible under the deterministic approach as the amount of data requested by the detailed knowledge of urban landscape is operationally unsuitable for hydrological modeling. The layout of the artificial network is derived from the available cartography and is shown - in the case of the selected portion of the city of Genova - in Figure 2a and 2b for comparison with both the natural and the underground networks.

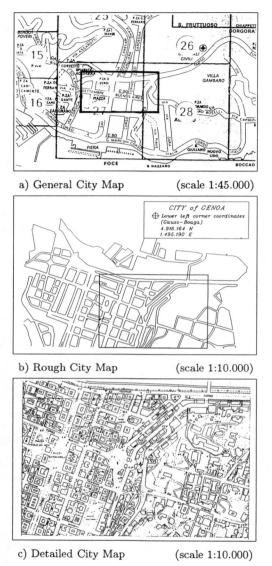

a) General City Map   (scale 1:45.000)

b) Rough City Map   (scale 1:10.000)

c) Detailed City Map   (scale 1:10.000)

Figure 1: Representation of the urban landscape features for a portion of the city of Genova reproduced at different resolution scales.

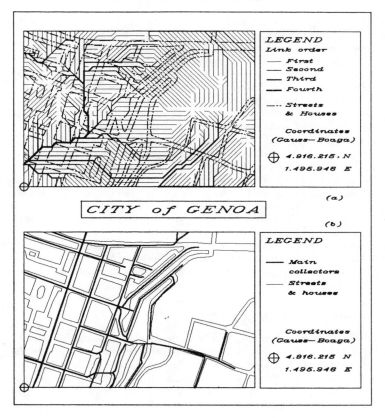

Figure 2: The natural, artificial and subsurface network for a portion of the urban landscape of the city of Genova.

The subsurface network is made up of pipes and channels - subdivided into main and secondary collectors - and follows the layout of the surface network, at least in the case of the main streets and pathways. The modeling of the flow routing within this network is not taken into account in the present analysis except for the identification of the variability of the inputs and the probability that connections between the surface and subsurface networks fail giving rise to local unefficiencies of the system. In Figure 2b the layout of the main collectors is outlined in order to show the incomplete correspondence between the surface and subsurface drainage system.

The major concept which underlies the present approach to the modeling of urban drainage systems is that of allowing a certain amount of water to move from one network to another as a function of the randomly positioned discontinuities as well as of the unefficiency of part of the connections. Once the water is not allowed to follow the design pattern, somewhere along the artificial network, not entering a gully hole for unefficiency reasons or due to some microscale effect, it is forced to choose - as a function of a suitable probability distribution - wether to continue its flowpath along the artificial network or to drain into the underlying natural system until it reaches some downward connection, somewhere else in the network. This is the reason assumed here as the main responsible for the occurrence of local flooding in unforeseen areas. In traditional urban hydrology just the first part of this concept is taken into account, i.e. the unefficiency of the gully holes, even if related to the discharge conditions of the underground network only. The stochastic distribution of microscale perturbations is the key factor which allows in this respect to produce different network scenarios for the simulated drainage system.

The possible connections between the different drainage networks are schematically described in Figure 3, which has been reproduced and modified from the scheme presented by Smith [8] when deling with the applications of distributed models - based on the data handling capabilities of Geographical Information Systems - in urban hydrology.

THE SIMULATION PROCEDURE

Within a stochastic framework the main operational steps envisaged as the basic skeleton of the proposed simulation procedure are described in the present section. A preliminary item is however related to the acquisition of the underlying information about landscape morphology at a suitable resolution scale. As a first approximation the latter may be assumed as the scale of the traditional urban drainage design. The selection of the minimum scale of representation of the landscape - e.g. the scale of the single house - is the first operational step to be considered. Down to this scale the recursive reproduction of the structural features of urban landscape is performed by means of a suitable self-affinity investigation.

Next, the definition of the probability distribution of gully holes location and efficiency is requested on the basis of some statistical analysis of available historical data. A suitable definition is also needed for the probability distribution of microscale effects which are prone to produce deviations from the expected drainage patterns. In this way a number of gully holes, randomly positioned along the network, will be charged with a drainage load which is larger than the design one. The definition is calibrated, in this case, on the basis of statistical sampling over the investigated urban domain.

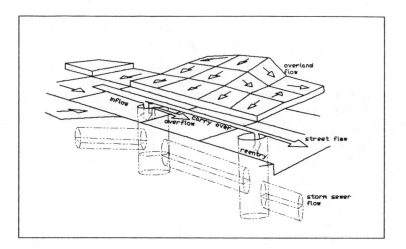

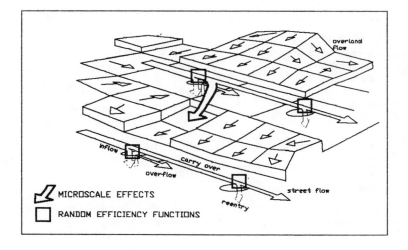

Figure 3: Scheme of possible connections between the networks in both the traditional (from Smith [8]) and the stochastic approach (modified from Smith [8]).

The recursive stochastic simulation - obtained by means of a series of generated microscale scenarios for any given rainfall event with the associated return periods - is the last item of the procedure. This would produce results which may be synthetized in the form of Table 1. Note that the difference between the deterministic output of the traditional approach and the expected value of the results obtained by the whole series of stochastic realizations might not be negligible for the randomness associated with microscale perturbations leads to the enhancement of critical effects. A further important conclusion to be drawn from the analysis of simulation results is related to the fact that the spatial identification of system failures as predicted from the stochastic approach is not likely to be consistent with the deterministic result showing the actual drainage efficiency at the large scale.

Table 1: Schematic representation of expected simulation results.

Rainfall event (return period)	T_1	T_2	T_3	T_4	T_5
N° of failures (traditional approach)	N_1	N_2	N_3	N_4	N_5
N° of failures (stochastic approach)					
realization n° 1	$n_{1,1}$	$n_{2,1}$	$n_{3,1}$	$n_{4,1}$	$n_{5,1}$
realization n° 2	$n_{1,2}$	$n_{2,2}$	$n_{3,2}$	$n_{4,2}$	$n_{5,2}$
realization n° 3	$n_{1,3}$	$n_{2,3}$	$n_{3,3}$	$n_{4,3}$	$n_{5,3}$
realization n° 4	$n_{1,4}$	$n_{2,4}$	$n_{3,4}$	$n_{4,4}$	$n_{5,4}$
realization n° ...	...	...	...	...	...
Expected value	E_1	E_2	E_3	E_4	E_5
C.V.	cv_1	cv_2	cv_3	cv_4	cv_5

CONCLUSIONS

A stochastic approach to the modeling of urban drainage networks has been presented in this paper with the aim of addressing the problem of the evaluation of the actual drainage efficiency in preventing local flooding due to microscale perturbations. The proposed methodology is based on the identification of the optimum resolution scale which is suitable for urban runoff modeling as derived from the analysis of the available data sets and computational capabilities. It has been recognized that the scale of micro perturbations is always finer than any resolution scale suitable for the deterministic description of urban landscape. After reconstructing the urban morphology down to the scale which is consistent with the identification of self-affinity characteristics - reproduced at different resolution scales within the urban structure - the stochastic modeling of microscale perturbations has been proposed. At the same time the probabilistic approach to the modeling of gully holes efficiency and location is envisaged for producing different urban scenarios where the simulation of rainfall-runoff processes - based on traditional design rainfalls for the input - is to be performed.

The main objective of the present work is that of pointing out the role played by microscale structures in determining deviations of the runoff flow from the design patterns as defined by the layout of streets and pathways making up the artificial drainage network traditionally simulated within a deterministic approach. This leads to a better understanding of the reasons for the occurrence of local flooding as usually experienced in highly urbanized areas as a consequence of heavy precipitation events, even if not exceeding the actual design rainfall. The traditional assumption of considering a fixed contributing area for each of the simulated gully holes - deterministically located along the artificial network - is thus herewith by passed allowing for statistically varying contributing areas to be drained by randomly positioned gully holes.

In this view the described methodology is not proposed as a comprehensive approach to urban drainage design but, much more likely, as a suitable procedure for the assessment of the actual drainage network efficiency and a tool for the identification of urban areas which are prone to be flooded during heavy rainfall events. This would be helpful, again, for the statistical evaluation of damages produced by local flooding to specific urban targets for which the actual distribution, all over the landscape, is not deterministically known at the scale of drainage modeling.

REFERENCES

1. Chow, V.T., Maidment, D.R. and Mays L.W. *Applied Hydrology*. Mc. Graw-Hill, 1988.
2. Huff, F.A. Time Distribution of Rainfall in Heavy Storms. *Water Resour. Res.*, 3(4), 1007 -1019, 1967.
3. Pilgrim, D.H. Bridging the Gap Between Flood Research and Design Practice. *Water Resour. Res.*, 22(9), supp., 165S-176S, 1986.
4. Linsley, R.K. Flood Estimates: How Good are They?. *Water Resour. Res.*, 22(9), supp., 159S-164S, 1986.
5. Watt, W.E. and Kidd, C.H.R. QUURM - A Realistic Urban Runoff Model. *J. Hydrology*, 27, 225-235, 1975.
6. Voorhees, M.L. and Wenzel, Jr.H.G. Urban Design-storm Sensitivity and Reliability. *J. Hydrology*, 68, 39-60, 1984.
7. Sarma, P.B.S., Delleur, J.W. and Rao, A.R. Comparison of Rainfall-runoff Models for Urban Areas. *J. Hydrology*, 18, 329-347, 1973.
8. Smith, M.B. A GIS-based Distributed Parameter Hydrologic Model for Urban Areas. *Hydrological Processes*, 7, 45-61, 1993.
9. Ando, Y., Musiake, K. and Takahasi, Y. Modeling of Hydrologic Processes in a Small Urbanized Hillslope Basin With Comments on the Effects of Urbanization. *J. Hydrology*, 68, 61-83, 1984.
10. Djokic, D. and Maidment D.R. Terrain Analysis for Urban Stormwater Modelling. *Hydrological Processes*, 5, 115-124, 1991.
11. Hollis, G.E. and Ovenden, J.C. The Quantity of Stormwater Runoff from Ten Stretches of Road, A Car Park and Eight Roofs in Herfordshire, England During 1983. *Hydrological Processes*, 2(3), 227-244, 1988.
12. O'Callaghan, J.F. and Mark, D.M. The Extraction of Drainage Network from Digital Elevation Data. *Comput. Vision Graph. and Image Proc.*, 28, 323-344, 1984.
13. Band, L.E. Topographic Partition of Watersheads with Digital Elevation Models. *Water Resour. Res.*, 22, 15-24, 1986.
14. La Barbera, P., Lanza, L. and Siccardi, F. Hydrologically Oriented Geographical Information Systems and Application in Rainfall-runoff Distributed Modeling: Case Study of the Arno Basin. In: it Application of Geographical Information Systems in Hydrology and Water Resources Management, ed. by K. Kovar and H.P. Nacthnebel, IAHS Pub. No. 211, 171-179, IAHS Press, Wallingford, UK, 1993.

Modelling of Flood Wave Propagation Over Flat Dry Areas of Complex Topography in Presence of Different Infrastructures

P. Molinaro, A. Di Filippo, F. Ferrari
ENEL-DSR-CRIS
Via Ornato 90/14
20162 Milan ITALY

ABSTRACT

After a description of the general equations which govern the propagation of a 2D flood wave (shallow water equations), a review of the scientific pubblications on the integration of their complete form is made, together with considerations on the adopted algorithms.
2D simplified models are successively considered. The aim of these models is mainly to deal with areas of complex topography; in this context a numerical model developed by ENEL-CRIS is described and its capabilities to take into account water behaviour in presence of hydraulic singularities and urbain areas are presented.

INTRODUCTION

One dimensional models are undoubtedly adequate for simulating flood propagation if the interested zone is a valley long and deep; this is frequently the situation for Italian mountain rivers.
There are however cases where one-dimensional models are not applicable, since it is not possible to define a valley axis a priori, so that the water velocity can be considered practically uniform through the valley cross sections. These cases are typical of very wide flat valleys, where there are urban areas or large infrastructures due to socioeconomic development, such as bridges, crossroads, and road or rail embankments. In case of exceptional flows, these constructions become obstacles to the water movement and cause backwater and local flooding. Two-dimensional flood models have become then an essential tool for locally refining the results produced by one-dimensional propagation models.

The general equations which govern the propagation of a 2D flood wave, generally called "shallow water equations", are as follows:

$$\frac{\partial h}{\partial t} + \frac{\partial q_x}{\partial x} + \frac{\partial q_y}{\partial y} = 0$$

$$\frac{\partial q_x}{\partial t} + u\frac{\partial q_x}{\partial x} + v\frac{\partial q_x}{\partial y} + u\left(\frac{\partial q_x}{\partial x} + \frac{\partial q_y}{\partial y}\right) + \left(gh - u^2\right)\frac{\partial h}{\partial x} - uv\frac{\partial h}{\partial y} + gh\left(C_f q_x - S_x\right) +$$

$$-\frac{\partial T_{xx}}{\partial x} - \frac{\partial T_{xy}}{\partial y} = 0$$

$$\frac{\partial q_y}{\partial t} + u\frac{\partial q_y}{\partial x} + v\frac{\partial q_y}{\partial y} + v\left(\frac{\partial q_x}{\partial x} + \frac{\partial q_y}{\partial y}\right) - uv\frac{\partial h}{\partial x} + \left(gh - v^2\right)\frac{\partial h}{\partial y} + gh\left(C_f q_y - S_y\right) +$$

$$-\frac{\partial T_{yx}}{\partial x} - \frac{\partial T_{yy}}{\partial y} = 0$$

where:

x, y = orthogonal coordinates in the horizontal plane;
t = time;
h = water depth;
u, v = components of the flow velocity vector in the x and y directions
q_x, q_y = specific flow rates in the x and y directions;
g = gravitational acceleration;
C_f = resistance coefficient;
S_x, S_y = ground slope components in the x and y directions;
$T_{xx}, T_{yx}, T_{xy}, T_{yy}$ = components of tensor of internal forces due to turbulence.

The resistance coefficient C_f is usually calculated according to the Manning formula:

$$C_f = \frac{n^2 |q|}{h^{\frac{10}{3}}} \qquad \text{where:} \qquad |q| = \sqrt{q_x^2 + q_y^2}$$
$$n = \text{Manning coeff}.$$

The terms T_{xx}, T_{yx}, T_{xy}, T_{yy} are commonly approximated by modelling them according to the following expressions:

$$T_{xx} = 2 \upsilon_t h \frac{\partial u}{\partial x}$$

$$T_{xy} = T_{yx} = \upsilon_t h \left(\frac{\partial u}{\partial y} + \frac{\partial v}{\partial x} \right)$$

$$T_{yy} = 2 \upsilon_t h \frac{\partial v}{\partial y}$$

Where υ_t is the "eddy viscosity" introduced by Boussinesq.
These terms are usually disregarded in flood models.
After a general overview of two dimensional flood models, this article describes more in detail the model FLOOD2D, developed by ENEL-CRIS to deal with 2D flood studies, and the algorithms successively added to allow its application in case of particularly complex terrain topography, with infrastructures, hydraulic singularities and urban areas.

2D MODELS BASED ON THE COMPLETE EQUATIONS

The scientific publications are full of solutions of the complete "shallow water" equations. Both the finite difference and finite-element methods are generally applied. A recent review of finite difference methods has been put together by Casulli [1], whereas a description of the finite-element method can be found in Pironneau [2] and Agoshkov et al. [3]. The majority of these methods are valid for simulating flows on regular ground with Froude number sufficiently less than one; otherwise, they run up against serious numerical problems.
In case of a flood wave the flow regime can change in space and time. A recent review of two-dimensional mathematical models of dam failure waves has been written by Ech-Cherif El Kettani et al. [4]. These authors propose a mathematical model based on the MacCormack scheme generalized to cases of non-Cartesian grids. The same authors have checked their models by comparing them with experimental data reported

by Bellos et al. [5] with good results.
Other researchers have also obtained excellent results using the MacCormack scheme, such as, for example, Jiménez and Chaudhry [6], and Murty Bhallamudi and Chaudhry [7].
A theoretical analysis of the characteristics of the MacCormack model has been presented by Di Monaco and Molinaro [8].
As an alternative to the MacCormack scheme, which is explicit and involves second-order approximations, other models have been used with greater or lesser success. On the subject of implicit models, one of the most important contributions is the Fennema and Chaudhry model [9]. These authors described factored implicit models, developed by Beam and Warming, for the conservative numerical solution of hyperbolic systems to which the shallow water equations belong. These implicit models are not iterative and therefore lead to a considerable time saving compared with others of the same type, especially for multi-dimensional problems. The formulations of the majority of these models have second-order accuracy in time and can be given second or fourth-order accuracy in space. In order to allow calculations in the presence of subcritical and supercritical flows, a "switching" technique is introduced to achieve appropriate space differentiation in a similar way to that of the "flux splitting" technique. These models have been applied with success by Fennema and Chaudhry themselves in the case of a reservoir emptying due to a dam collapsing. Examples have been studied with a small ratio between water depths upstream and downstream of the dam; however, due to the dissipative character of the algorithm, the solution turns out to be smoothed above a certain number of grid points if steep fronts are formed.
A finite-element model for simulating subcritical and supercritical flows has been proposed by Di Monaco and Molinaro [10]. In this model, the water depth is approximated with first-order functions and the velocity components are integrated over depth with second order functions; this choice prevents the formation of parasitic waves in the numerical solution. The time integration is carried out using a one-step implicit method. This model has been tested in the cases of reservoir emptying and propagation on dry bed, with good results. In both cases it was necessary to use an appropriate value for artificial viscosity in order to achieve convergence during the iterative solution process.
Another approach to finite elements has been proposed by Hervouet [11]. This model makes use of the fractional step method defined as follows: a) first, the velocity components are updated using the convective terms only; b) then, the so-called propagation-diffusion problem is solved. Bilinear functions are used for all dependent variables. The author shows the ability of the model to simulate flow regime transitions and flooding

and drying of the ground in a series of cases of practical interest.

Among the finite-element models which are able to describe shock waves, there is one proposed by Katopodes [12]. This model follows the Petrov-Galerkin approach, according to which the weight functions are a suitable combination of the shape functions and of their first space derivatives.

A recent analysis of finite-element models has been presented by Agoshkov et al. [13], who propose different alternative implicit or semi-implicit formulations. The authors themselves analysed the various conservative and non-conservative forms of the basic equations in detail, along with the respective boundary conditions which naturally derive from the integral formulation of the so-called "weak" problem.

Among the most recent models worthy of note there is the Priestley's model [14], which is based on the "flux-splitting" technique, applied with different flux limiters, such as those of Van Leer and Roe. This model uses Cartesian or curvilinear grids. The Cartesian grids can be closed with oblique sides in order to follow better the domain boundary.

Flux-splitting methods, are accurate both in zones where the flow changes gradually and where there are steep profiles, such as hydraulic jumps and shock waves.

As far as the writers know, up to now the ability of complete models to simulate the propagation of flood waves on dry beds has only been shown in cases of flat or slightly irregular ground. This explains why simplified models have had great success in practice to study the propagation of floods on highly irregular beds.

The flooding and drying process of flat areas has been simulated successfully with complete models, when there are slow changes in flow characteristics, by introducing a check on the depth which more or less works as follows:

- if the depth of the water is greater than a certain tolerance, the standard equations are applied in order to calculate the updated values of the flow variables;
- if the depth of the water is less than the same tolerance, then the velocity components are set to zero and the depth value is redefined and set equal to the tolerance.

As many authors have reported, this trick works well both for finite difference and finite-element models.

2D SIMPLIFIED MODELS

The basic equations can be simplified by eliminating the inertial terms. This approximation is strictly only admissible when the Froude number is much smaller than one; this case can occur when the area under study has a relatively gentle slope.

The first generation of simplified models is that of the so-called quasi-two-dimensional models, based on various interconnected cells, such as Zanobetti and Lorgeré Mekong delta model [15]. This model assumes that flooding takes place quite slowly, so that it is legitimate to neglect he inertial terms; consequently, the exchange flow rate between cells lepends solely on the water level difference through a suitable functional relationship.

The successive generation of models is different from the first one since the calculation domain is represented as a continuum, and the velocity components are calculated using momentum equations between two contiguous cells. All inertial terms are neglected in the momentum equations, which gives rise to the so-called parabolic approximation; however eliminating the flow rate time derivatives does not seem to be necessary and does not lead to any significant computational advantage.

Todini et al. [16] and Reitano [17] use a grid of triangles of variable size which allows the model to be well adapted to the shape of the river bed as well as to the topographic characteristics of the adjacent areas.

The model developed by Hromadka and Yen, and re-proposed by Maione et al. [18], has the interesting characteristic of using the continuity condition to couple an explicit two- dimensional model, solved on a Cartesian grid, with a one- dimensional model, also based on the diffusive wave approximation.

The most recent generation of simplified models are obtained by neglecting the convective terms only; the basic equations therefore become:

$$\frac{\partial h}{\partial t} + \frac{\partial q_x}{\partial x} + \frac{\partial q_y}{\partial y} = 0$$

$$\frac{\partial q_x}{\partial t} + gh\frac{\partial h}{\partial x} + gh\left(C_f q_x - S_x\right) = 0$$

$$\frac{\partial q_y}{\partial t} + gh\frac{\partial h}{\partial y} + gh\left(C_f q_y - S_y\right) = 0$$

From a practical point of view, using these equations even in cases where the conditions for their validity do not entirely occur can be justified by the following argument: kinetic energy changes are generally only important in relatively small areas in comparison with the whole flooding domain, so that the simplified equations give a good approximation of reality over most of the domain itself; furthermore, approximate topographic representation of the bed and roughness can lead to errors in the results which are even greater than those caused by using simplified equations. This does not mean that models which use complete two-dimensional equations are devoid of interest, since they represent, in any case, the modelling level which should be attained if possible.

One possible scheme to discretize in time the simplified equations is as follows:

$$
\begin{cases}
\dfrac{h^{n+1} - h^n}{\Delta t} + \left(\dfrac{\partial q_x}{\partial x}\right)^{n+1} + \left(\dfrac{\partial q_y}{\partial y}\right)^{n+1} = 0 \\[3mm]
\dfrac{q_x^{n+1} - q_x^n}{\Delta t} + gh^n\left(\dfrac{\partial h}{\partial x}\right)^{n+\theta} + gh^n\left(C_f^n q_x^{n+1} - S_x\right) = 0 \\[3mm]
\dfrac{q_y^{n+1} - q_y^n}{\Delta t} + gh^n\left(\dfrac{\partial h}{\partial y}\right)^{n+\theta} + gh^n\left(C_f^n q_y^{n+1} - S_y\right) = 0
\end{cases}
$$

Where the parameter "θ" can take values from 0 to 1. $\theta = 0$ gives an explicit model in terms of water depth gradients, which is easy to set up, but has a limited time integration step; $\theta = 1$ gives a completely implicit model, which is much more complicated to implement, but, however, allows large time steps.
The space derivatives of the specific flow rates which appear in the continuity equation are calculated at time "n+1", since the model is stable for all values of θ only in this case, as can be seen from the analysis shown in [19].

ENEL-CRIS has developed a computer program called FLOOD2D based on the previous model. The solution algorithm can be both implicit ($\theta=1$) and explicit ($\theta=0$).

For the space discretization it is used an orthogonal staggered grid as the one shown in the following figure:

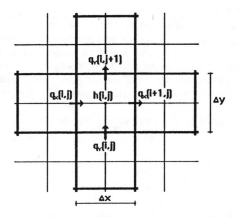

From the above equations the following expressions for the flow rate q_x and q_y at time "n+1" are obtained:

$$q_x^{n+1} = -\frac{gh^n\Delta t}{1+gh^nC_f^n\Delta t}\left(\frac{\partial h}{\partial x}\right)^{n+\theta} + \frac{gh^nS_x\Delta t+q_x^n}{1+gh^nC_f^n\Delta t}$$

$$q_y^{n+1} = -\frac{gh^n\Delta t}{1+gh^nC_f^n\Delta t}\left(\frac{\partial h}{\partial y}\right)^{n+\theta} + \frac{gh^nS_y\Delta t+q_y^n}{1+gh^nC_f^n\Delta t}$$

The model equations are solved for $\theta \neq 0$ with the iterative method described below:

1) attempt values are assumed for $h^{n+\theta}$; if h^p is the pth attempt value for $h^{n+\theta}$, then:
$\quad h^p = h^n \quad$ for $\quad p = 0$;

2) the corresponding values of q_x^{n+1} and q_y^{n+1} at the pth attempt, which are denoted as q_x^p and q_y^p, are calculated by substituting h^p to $h^{n+\theta}$ in the previous equations;

3) the residual is calculated by substituting the values of h^p, q_x^p and q_y^p into the continuity equation:

$$R^p = \frac{h^p - h^n}{\Delta t} + \frac{\partial q_x^p}{\partial x} + \frac{\partial q_y^p}{\partial y}$$

4) the correction Δh^p to be added to h^p so that $R\ (h^p + \Delta h^p) = 0$ is calculated; this correction is given by the following equation (see [19]):

$$\frac{\Delta h^p}{\Delta t} - \frac{\Delta\,tgh^n}{1 + \Delta\,tgh^nC_f^n}\frac{\partial^2\Delta h^p}{\partial x^2} - \frac{\Delta\,tgh^n}{1 + \Delta\,tgh^nC_f^n}\frac{\partial^2\Delta h^p}{\partial y^2} = -R^p$$

5) the correction is made:

$$h^{p+1} = h^p + \Delta h^p$$

6) the convergence test is carried out:

$$\left|\Delta h^p_{\max}\right| \le \varepsilon$$

If the above condition is not satisfied, the process returns to point 2, otherwise the iterative process is stopped.

Particular attention is paid to the interpolation process of the height of the water passing through the sides between the calculation cells; this is done in order to simulate incipient flood processes, embankment overflows or simple water passage on a sloping soil.

The bottom elevation between two adjacent cells is computed according to the following procedure.
A control parameter "VMRH" is introduced, which represents an upper bound for the ratio between the absolute value of the difference of the water depths at two adjacent cells "DH" and their mean value "HN".
First this ratio, which we refer to as "DHHN", is computed; its value is compared with the value of the control parameter "VMRH".

The following cases are possible:

case 1) $DHHN \ge VMRH$;
the difference between water depths is large because of the strong irregularity of the ground; the bottom elevation at the interface is computed as:
$$ZFFON = ZFMAX = \max\left[ZF(I,J), ZF(I-1,J)\right];$$

case 2) DHHN < VMRH;
the water depth does not change much from cell to cell and the bottom
elevation is computed according to the following expression:

$$ZFFON = \left(1 - \frac{DHHN}{VMRH}\right) * ZFMED + \left(\frac{DHHN}{VMRH}\right) * ZFMAX \quad .$$

The above formula ensures the continuity of the bottom elevation when a
transition occurs between case1 and case 2. The meaning of each term
can be better understood by means of the following scheme:

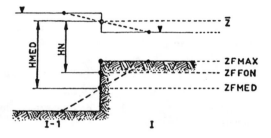

Making reference to the above scheme where the meaning of each term is
illustrated, if "H" is the water depth in correspondence of the center of a
cell, then the mean water depth is:

$$HMED = 0,5 * \left[H(I,J) + H(I-1,J) \right] \; ;$$

while the average free surface elevation is:

$$\overline{Z} = ZFMED + HMED \quad .$$

The water depth to be considered between two adjacent cells will result:

$$HN = \overline{Z} - ZFFON \quad .$$

If the value of HN is less or equal zero then the discharge is set to zero.
From practical applications "VMRH" is given values ranging from 0,25
to 0,50.
The different possible situations considered by the code are reported in
figure 1.

One important characteristic of the FLOOD2D program is its ability to adapt the calculation domain to the propagation of the flood. In particular, during the expansion phase, a new set of cells surrounding the calculation domain is added automatically after each time step; on the contrary the cells are eliminated if the flood recedes. This characteristic allows CPU time to be considerably reduced and saves the user from the definition of the domain extension inside the calculation rectangle.

Some results of FLOOD2D program applications relating to different topographical situations are presented graphically.
In particular, figure 2 refers to an initially dry very steep valley which is flooded from a tributary on the right slope; the water goes upstream as far as 4 Km from the section of the tributary.
Figure 3 refers instead to the branches of a river flowing in a lagoon with emerging land and deep channels; the distribution of the flow rates and the water circulation are effectively simulated by the model.

Other authors, such as Gallati et al. [20] and Natale et al. [21], have developed flood models from the same simplified expressions of the general shallow water equations; they also used Cartesian calculation grids, but with a scheme explicit in time.

All the two-dimensional models presented above are suitable for simulating the flooding of vast flat areas caused by bank overflow or backwater effects. However, it will not go unnoticed that so far nothing has been said yet about how to deal with flow through constructions on the river bed, such as bridges or crossings, or in urban areas: situations which are not infrequently met in reality. In these cases, the general equations are not valid as they are, and it is necessary to make modifications or use other equations.

TREATMENT OF HYDRAULIC SINGULARITIES

To study water flow through structures built on the river bed or on the flooded land the geometrical description of each construction should be as accurate as possible. An adequate algorithm should be able to deal with different kinds of flows such as: pressure flow, weir flow, free flow, submerged flow or flow with different kinds of orifices.
A recent development of suitable equations for dealing with such situations has been carried on by P. Molinaro and R. Pacheco [22], and implemented in the code FLOOD2D. The following procedure has been applied. A generic equation for computing the flow through a singularity may be written as follows (see also figure below):

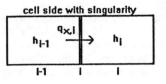

$$q_{x,i} = f_i (h_{i-1}; h_i)$$

where:

h_i = water depth on the cell "i";
$q_{x,i}$ = water discharge per unit width on the cell side.

The correction of the water discharge at a certain iteration "p" can be written as a function of water depth variations:

$$\Delta q_{x,i} = \left(\frac{\partial f_i}{\partial h_{i-1}}\right)^p \Delta h_{i-1} + \left(\frac{\partial f_i}{\partial h_i}\right)^p \Delta h_i$$

The above expression is used in the iterative process for the calculation of water depth corrections Δh^p described in the previous paragraph.
In the next figure an example of singularity is reported as it can be treated by FLOOD2D code.

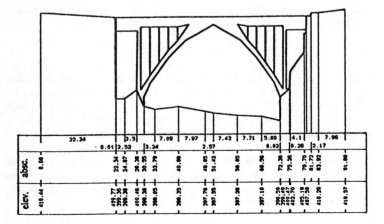

TREATMENT OF URBAN AREAS

For the flooding of urban areas, two alternative approaches are possible. The first, most detailed, is to use a channel network type model; but this solution lies outside the subject of this chapter: interested readers should refer to reference [23]. The second approach is to keep a two-dimensional representation of the flood and take into account that only part of the total built-up area is available for water storage or water flow. One of the first attempts to represent this two-dimensional situation was made by Braschi and Gallati [24], who introduced the concepts of "urban porosity" and "transmissivity" of a built- up area, with clear analogy to flow in a porous fractured medium.

The same concept was used by ENEL-CRIS to allow FLOOD2D to deal with flooding of urban areas.

Considering a particular cell of the grid of computation over an inhabited zone, the total area of the cell that can be flooded is generally reduced by the buildings whose area is not available for water storage.

In fact, looking at the following scheme where the buildings inside a computational cell are drawn in black, it is possible to say that the effective area "A_e" that can store water is:

$$A_e = n_l \, \Delta x \, \Delta y + n_l \, \Delta y (1 - n_l) \, \Delta x$$

where:

n_l = ratio between the "free length" and the total length (Δx or Δy) of each cell side;

$\Delta x, \Delta y$ = length of the cell side.

From the above relationship it is possible to evaluate the ratio "n_A" between the effective area and the total area as a function of "n_l":

$$n_A = \frac{A_e}{\Delta x \, \Delta y} = n_l + n_l (1 - n_l) = n_l (2 - n_l)$$

The new parameters "n_A" and "n_l" may be referred to as "areal porosity" and "linear porosity" of a built-up area.

The parameter "n_l" can be defined as a function of "i, j" indexes.

The continuity equation should than be modified considering the effective area of a cell and the "free length" of each side; following this line the general discretized form of this equation becomes, with reference to a generic "i, j" cell:

$$n_l(2-n_l)\frac{dh}{dt} + \frac{\tilde{n}_{l,i+1}\, q_{x,i+1} - \tilde{n}_{l,i}\, q_{x,i}}{\Delta x_i} + \frac{\tilde{n}_{l,j+1}\, q_{y,j+1} - \tilde{n}_{l,j}\, q_{y,j}}{\Delta y_j} = 0$$

where:

$$\tilde{n}_{l,i} = \min\left(n_{l,i}\,;\, n_{l,i+1}\right) \qquad\qquad \tilde{n}_{l,j} = \min\left(n_{l,j}\,;\, n_{l,j+1}\right)$$

In the above expression "i" and "j" indexes are omitted, for the sake of simplicity, when they are superfluous.

The above expression is implicitly based on the hypothesis that water can not get into buildings, and this is approximately true during the phase of fast level growth; subsequently, even if water really get into buildings, time level variations are generally more moderate so that the storage term is smaller with respect to those representing the exchange of water between adjacent cells.

CONCLUSIONS

A general description of the main models presently in use for the simulation in two-dimensions of flood propagation in flat areas has been given.

The computer code FLOOD2D developed by ENEL-CRIS has been presented. This code is based on a simplified version of the shallow water equations, solved by means of the finite volume technique applied on a cartesian grid. The code allows the simulation of flooding and drying processes in areas of complex topography as well as the computation of flow through hydraulic structures and urban areas.

Present development is addressed to interfacing FLOOD2D with a Geographycal Information System and Remote Sensing techniques, in order to have detailed and updated representation of the flood-plains topographycal and physical characteristics. Another improvement will be the introduction in the momentum equations of the convective terms.

REFERENCES

1. Casulli, V. Numerical simulation of shallow water flow. Computational Methods in Surface Hydrology. (Ed. Gambolati, G. , Rinaldo, A. , Brebbia, C.A. , Gray, W. G. , Pinder G. F.), pp. 13-22 , Proceedings of the 8th Int. Conf. on Computational Methods in Water Resources, Venice, Italy, 1990. Computational Mechanics Publications 1990.

2. Pironneau, O. Finite Element Methods For Fluids. John Wiley & Sons, Chichester and New York, 1989.

3. Agoshkov, V. I. , Ambrosi D. , Pennati, V. , Quarteroni, A. , Saleri, F. Mathematical and numerical modelling of shallow water flow. Computational Mechanics, 1993.

4. Ech-Cherif El Kettani, B. , Berrada, A. , Ouazar, D. , Agouzoul, M. 2D dam-break flood-wawe propagation on dry beds. Computer Methods and Water Resources II (Ed. Ben Sari, D. , Brebbia, C. A. , Ouazar, D.) , Proceedings of the 2nd Int. Conf. on Computer Methods and Water Resources, Rabat, Morocco, 1991. Computational Mechanics Publications 1991.

5. Bellos, C. V. , Soulis, J. V. , Sakkas , J. G. Experimental investigation of two-dimensional dam-break induced flows. Journal of Hydraulic Research, Vol. 30, N. 1, pp. 47-64 , 1992.

6. Jiménez, O. F. , Chaudhry, M. H. Computation of supercritical free-surface flows. Journal of Hydraulic Engineering, Vol. 114, N. 4, pp. 377-395, 1988.

7. Murty Bhallamudi, S. , Chaudhry, M. H. Computation of flows in open channel transitions. Journal of Hydraulic Research, Vol. 30, N. 1, pp. 77-94 , 1992.

8. Di Monaco, A. , Molinaro, P. Discussion of the paper by Bellos, C. V. , Sakkas, J. G.: 1-D Dam-break flood-wave propagation on dry bed. ASCE, Journal of Hydraulic Engineering, Vol. 115, N. 8, 1989.

9. Fennema, R. J. , Chaudhry, M. H. Implicit methods for two dimensional unsteady free-surface flows. Journal of Hydraulic Research, Vol. 27, N. 3, pp. 321-332 , 1989.

10. Di Monaco, A. , Molinaro, P. A finite element two-dimensional model of free-surface flows: verification against experimental data for the problem of the emptying of a reservoir due to dam-breaking. Computer Methods and Water Resources (Ed. Ouazar, D. , Brebbia, C. A. , Barthet, H.) , Proceedings of the 1st Int. Conf. on Computer Methods and Water Resources, Rabat, Morocco, 1988. Computational Mechanics Publications 1991.

11. Hervouet, J. M. Telemac, a fully vectorised finite element software for shallow water equations. Computer Methods and Water Resources II (Ed. Ben Sari, D. , Brebbia, C. A., Ouazar, D.) , Proceedings of the 2nd Int. Conf. on Computer Methods and Water Resources, Rabat, Morocco, 1991. Computational Mechanics Publications 1991.

12. Katopodes, N. D. A dissipative Galerkin scheme for open-channel flows. Journal of Hydraulic Engineering, Vol. 110 , 1984.

13. Agoshkov, V. I. , Ovchinnikov, E. , Quarteroni, A. , Saleri, F. Recent developments in the numerical simulation of shallow water equations. II. Temporal discretization. Mathematical Models and Methods in Applied Sciences.

14. Priestley, A. New Numerical Algorithms for Conservation Laws. Final Report for ENEL-CRIS, Milan, Italy, 1992.

15. Zanobetti, D. , Lorgeré, H. Le modèle mathematique du delta du Mekong. La Houille blanche, 1-4-5, pp. 17-20, 255-269, 363-378.

16. Todini, E. , Venutelli, M. Overland flow: a two-dimensional modelling approach. Atti Nato ASI, Workshop on Recent Advances in Hydrology and Water Resources, 1988.

17. Reitano, B. Modello bidimensionale per la simulazione di inondazioni fluviali. 23° Convegno di Idraulica e Costruzioni Idrauliche, Firenze, 1992.

18. Maione U. , Mignosa P. , Tanda M. G. Modello matematico per l'allagamento della Piana di Selvetta (Sondrio). Rapporto 1988 del G.N.D.C.I. , Linea 1.

19. Molinaro, P. , Di Filippo, A. , Ferrari, F. Un modello matematico per la simulazione delle inondazioni di vaste aree a topografia complessa:

aspetti teorici, informatici ed applicativi. Relazione ENEL-CRIS n. 4514, settembre 1992.

20. Gallati, M. , Braschi, G. , Di Filippo, A. , Rossi, U. Simulation of the inundation of large areas of complex topography caused by heavy floods. HYDROSOFT' 90 (Ed. Blain, W. R. , Ouazar, D.) , Proceedings Int. Conf. Hydraulic Engineering Software Applications, Boston, Massachussetts, U.S.A. , 1990, Computational Mechanics Publications 1990.

21. Natale, L. , Savi, F. Espansione di onde di sommersione su terreno inizialmente asciutto. Idrotecnica, 6, pp. 397-406, 1991.

22. Molinaro, P. , Pacheco, R. Sul calcolo della portata transitante attraverso singolarità presenti in un alveo naturale. Relazione ENEL-CRIS n. 4840, marzo 1994.

23. Braschi, G. , Gallati, M. , Natale, L. La simulazione delle inondazioni in ambiente urbano. C.N.R. , quaderni G.N.D.C.I. , 1990.

24. Braschi, G. , Gallati, M. Simulation of a levee-breaking submersion of planes and urban areas. HYDROCOMP '89, Proceedings International Conference on Computational Modelling and Experimental Methods in Hydraulics, Elsevier Applied Science, p.p. 117-126, 1989.

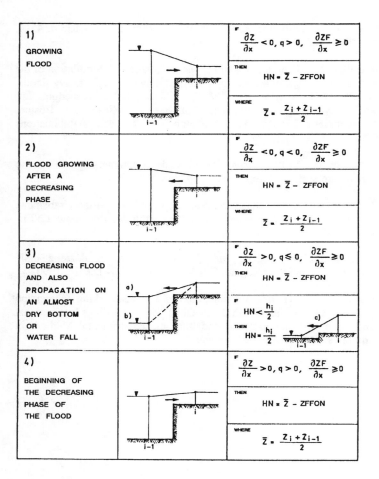

Fig.1 - Where:

Z	=	water level of the cell (equal to ZF+H);
ZF	=	bottom elevation of the cell;
$\overline{Z}$	=	average water level between two cells;
ZFFON	=	bottom elevation between two cells;
HN	=	water depth considered between two cells.

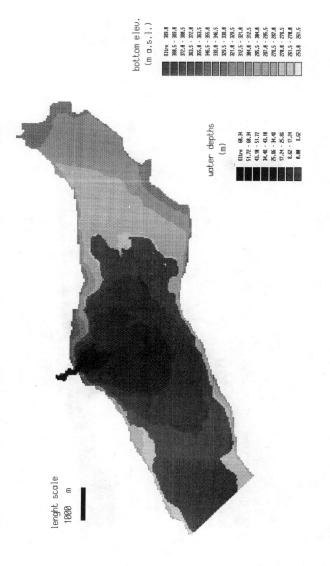

bottom elev.
(m a.s.l.)

	Oltre 389.0
	380.5 - 389.0
	372.0 - 380.5
	363.5 - 372.0
	355.0 - 363.5
	346.5 - 355.0
	338.0 - 346.5
	329.5 - 338.0
	321.0 - 329.5
	312.5 - 321.0
	304.0 - 312.5
	295.5 - 304.0
	287.0 - 295.5
	278.5 - 287.0
	270.0 - 278.5
	261.5 - 270.0
	253.0 - 261.5

water depths
(m)

	Oltre 60.34
	51.72 - 60.34
	43.10 - 51.72
	34.48 - 43.10
	25.86 - 34.48
	17.24 - 25.86
	8.62 - 17.24
	0.00 - 8.62

lenght scale
1000   m

Fig.2 - Wetted area during a flood event.

- 227 -

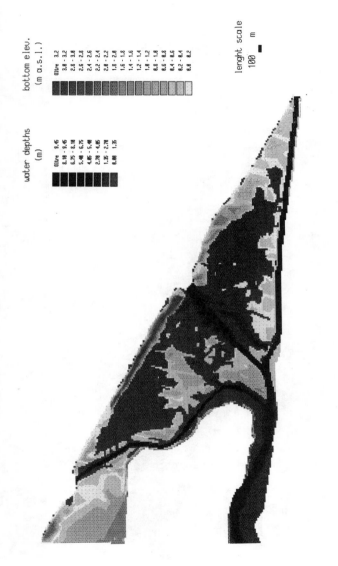

Fig.3 - Wetted area of a lagoon, with deep channels and emerging land during a flood event.

New Methods for Modelling Dam-Break Wave.

A. PAQUIER
CEMAGREF
3 bis, quai Chauveau
69336 LYON CEDEX 09
FRANCE

ABSTRACT

Recent developments in numerical analysis for hyperbolic equations have led to new methods which solve shallow water equations including complementary equations for hydraulic jumps. An explicit finite volume scheme has also been developed to solve 2-D problems such as flooding of an initially dry plain. The study of the failure of Agly River Dam implied the modelling of a dam-break wave coming over an embankment which crossed the flow direction.
In 1-D, sediment transport can also be modelled in a simplified way. The example of the historical failure of Lawn Lake Dam shows the influence of sediment deposits on the maximum water levels and, more generally, the interest of such a modelling.

INTRODUCTION

After Second World War, in France, government planned the construction of numerous dams for the production of electricity. This program was led by the newly created national firm EDF (Electricité de France) which developed important works for safety of dams. After Malpasset dam failure in 1959, decrees oblige owners of dams to set up emergency plans in order to organise quick evacuation of people living in the valley in case of dam-break. Modelling of dam-break wave was thus necessary to estimate the submerged areas and the time available for evacuation. Physical models were first used but since the sixties, development of mathematical models have occurred at EDF ([1]).
In the seventies, following the development of irrigation, more dams were constructed for agricultural purposes. In order to help the services of the Ministry of Agriculture, CEMAGREF, formerly CTGREF, the French Institute of Agricultural and Environmental Engineering Research, developed

researches as well in the field of the construction of dams as for their maintenance and safety and, particularly, computation of dam-break waves.

Dams for agricultural purposes were of varied importance so that 2 kinds of methods were developed for computation of dam-break waves :

- "simplified" methods which laid on a dimensionless graph. They were developed for small and medium dams which did not demand emergency plans. The last version is a software called CASTOR which can be used on a personal computer and requires a few seconds for computation.

- "complete" methods solving shallow water equations. Such methods are necessary for large dams (in France, when the dam is more than 20 metres high and the volume of the reservoir is more than 15 millions cubic metres). Usual computations are based on 1-D codes which historically were the first ones. Thanks to the increase of computer power, 2-D computations are now possible and used for particular situations.

For both 1-D and 2-D codes solving shallow water equations, the specific difficulties of computing dam-break waves are :

- the transitions from a supercritical flow to a subcritical one and inversely ;
- the progression of a front on initially dry land.

Both cases mean a discontinuity in some of the hydraulic variables. So, since 1984, for CEMAGREF, J. P. Vila has developed numerical methods which make possible computing discontinuous solutions of shallow water equations ([5, 6, 7]).

Here below, we shall first describe these methods when applied to 2-D models. We shall insist on the specific advantages of such methods thanks to the example of the Agly River Dam. In a second part, we shall mention the extensions to 1-D code which deal with sediment transport ; the example of the Lawn Lake Dam will show the interest of such a computation.

NUMERICAL METHOD FOR THE 2-D CODE RUBAR 20

Since 1988, CEMAGREF has developed the code called RUBAR 20 which solves the 2-D shallow water equations. These equations are written under the following conservative form :

$$\frac{\partial h}{\partial t} + \frac{\partial U}{\partial x} + \frac{\partial V}{\partial y} = 0 \tag{1}$$

$$\frac{\partial U}{\partial t} + \frac{\partial \left(\frac{U^2}{h} + g\frac{h^2}{2} \right)}{\partial x} + \frac{\partial \left(\frac{UV}{h} \right)}{\partial y} = -gh\frac{\partial Z}{\partial x} - g\frac{U\sqrt{U^2+V^2}}{C^2h^2} \tag{2}$$

$$\frac{\partial V}{\partial t} + \frac{\partial \left(\frac{UV}{h} \right)}{\partial x} + \frac{\partial \left(\frac{V^2}{h} + g\frac{h^2}{2} \right)}{\partial y} = -gh\frac{\partial Z}{\partial y} - g\frac{V\sqrt{U^2+V^2}}{C^2h^2} \tag{3}$$

where h is the water depth, Z the bed elevation, U and V the "discharges" (water depth multiplied by velocity) along respectively the x-axis and the y-axis and C the Chezy's friction coefficient which is considered as a constant or equalled to $K\,h^{1/6}$ where K is the Strickler's friction coefficient.

The second member of equations (1), (2) and (3) may be completed by including wind-effect, diffusion, etc.

The computational area is divided into an irregular mesh formed of quadrilaterals or triangles. 2 adjacent cells have one complete common edge. The finite volume scheme used is explicit and second-order accurate.

The solution of the system of equations (1) + (2) + (3) is obtained through 2 steps :

- computation of the fluxes through edges for the first member ;

- integration (on the surface of the cell) of values at the centre of the cell for the second member.

First member

Fluxes through one edge are computed by solving an homogeneous 1-D Riemann problem perpendicularly to the edge and then using the stationary value for the computation. As the system of equations (1) + (2) +(3) without first member is invariant by rotation, it remains unchanged in the reference system constituted by the normal and the tangent to any edge. If we neglect the propagation in the direction tangent to the edge, equations are simplified. At the middle of the edge, we thus solve the following equations :

$$\frac{\partial h}{\partial t} + \frac{\partial U}{\partial x} = 0 \qquad (4)$$

$$\frac{\partial U}{\partial t} + \frac{\partial\left(\dfrac{U^2}{h} + g\dfrac{h^2}{2}\right)}{\partial x} = 0 \qquad (5)$$

$$\frac{\partial V}{\partial t} + \frac{\partial\left(\dfrac{UV}{h}\right)}{\partial x} = 0 \qquad (6)$$

where the x-axis is perpendicular to the edge.

The equations (4) + (5) +(6) constitute a non-linear hyperbolic system for the 3-D variable $W(x, t) = (h, U, V)$. Eigenvalues are u-c, u, u+c where u is the velocity perpendicular to the edge and $c = \sqrt{gh}$ is the wave celerity. The second field is linearly degenerated whereas the first and the third ones are truly non-linear. The Riemann invariants can be very simply computed as they respectively equal u+2c and v, u+2c and u-2c, u-2c and v.

Riemann problem is approximately solved in such a way that the property of the fields are kept. In fig. 1, we show a particular example where the first field corresponds to a rarefaction and the third one to a shock. WAL is obtained from WL by conservation of the first Riemann invariants with limits of slopes u_L-c_L and u_A-c_A (exact solution). WAL and WAR are separated by a jump of v, i.e. from $(h_A, h_A\,u_A, h_A\,v_1)$ to $(h_A, h_A\,u_A, h_A\,v_2)$, which is also the exact solution. Then the speed of the 3-shock is approximated as it is computed as half the sum of the left (corresponding to WAR) and right (corresponding to WR) 3-eigenvalues.

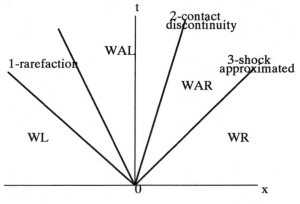

Fig.1-Solution of Riemann problem

In order to obtain a second-order scheme (in space), following ideas by B. Van Leer ([4]), the values considered in the Riemann problem are computed by using slopes in the x and y directions (for each of the 3 variables $z = h + Z$, U and V independently). For every cell, these slopes are computed by the method of the least squares (on each of the 3 variables independently) on the basis of the values at the centre of the cells which have one common edge with the given cell (it is possible to do it by adding the cells with 1 common summit). The slope for any variable is later limited if the value of the given variable A (which may be z, U or V) at the middle of one edge is not located between the value at the centre of the cell A_c and $\alpha A_n + (1-\alpha)A_c$ where A_n is the value at the centre of the corresponding adjacent cell. The parameter α should be selected between 0.5 and 1. Better results are generally obtained with α equalling 0.6.

In order to obtain second-order accuracy in time, the values of fluxes should be assessed at time $t_{n+1/2} = t_n + 0.5\ \Delta t$ instead of time t_n. Inside every cell, independently from other cells, thanks to the slopes previously computed, a first assessment of fluxes (at time t_n) is obtained which is used to compute the values of the variables at the middle of the edge at time $t_{n+1/2}$. The same calculation in the adjacent cell will give the other (left or right) values for the Riemann problem.

Second member
Using the fluxes across edges, it is possible to obtain provisional values of the variables at time $t_{n+1} = t_n + \Delta t$. We then add the contribution of the second member which includes, at least, 2 terms :

- gravity or slope terms $(-gh\dfrac{\partial Z}{\partial x}$ or $-gh\dfrac{\partial Z}{\partial y})$;

-bed friction by the use of a Chezy's coefficient (small water depths) or a Strickler's coefficient.

Gravity terms are treated as fluxes in such a way as an horizontal water surface remains strictly horizontal. For instance, the integral on one cell $\iint -gh\dfrac{\partial Z}{\partial x}dxdy$ becomes $0.5\sum\limits_{edges}\alpha_i\dfrac{L_i}{S}(h_c + h_i)(Z_i - Z_c)$ where α_i is the x-coordinate of the normal to the edge i, L_i is the length of the edge, h_c and Z_c respectively the water depth and the bed elevation at the centre of the cell, h_i and Z_i respectively the water depth and the bed elevation at the middle of the edge i, S the area of the cell.

Bed friction terms are more simply assessed at the centre of the cell by using the value at time t_n and the provisional value at time t_{n+1}. Possible additional terms of second member which are supposed of minor importance are generally computed in a similar way as bed friction, i.e. at the centre of the cell.

Properties of the scheme
About the numerical scheme here above described, we note the good conservation of volumes ; accuracy is about 10^{-4} in an ordinary computation of dam-break wave on initially dry areas. Moreover, the computation of discontinuities is included in the scheme as ordinary points so that there is no need of any particular treatment which is usually source of errors and difficulties. A front coming on initially dry land can be treated in the same way. Mathematically, the Riemann problem can be solved similarly when water depth is zero on one side. However, computer may give a value slightly different from 0 instead of 0 ; after a lot of time steps, this may lead to errors. We thus decide to choose minimum values (of water depth and velocity) below which variables are set back to 0. For usual problems of dam-break waves, these minimum values are 10^{-4} m for water depth and 10^{-3} m/s for velocity.

A CASE TREATED BY RUBAR 20 : AGLY RIVER DAM

The dam under construction on Agly River is located in the region of Languedoc - Roussillon in the South of France. It is about 60 metres high for a volume of the reservoir equalling 25 millions cubic metres at the normal water level. Dam-break wave was computed along the 30 kilometres of valley downstream the dam using the 1-D code RUBAR 3. The maximum discharge obtained downstream was about 12,000 m^3/s ([3]) whereas the discharge corresponding to the return period of 100 years was about 2,000 m^3/s. It thus appeared obvious that along the 15 kilometres of river further downstream towards Mediterranean Sea, the flow would pass over the embankments and inundate the plain. In that area, the plain of Salanque is rather flat and it is difficult to define favoured beds for the flow except the minor bed which would receive 10 to 15% the peak discharge. 2-D computation was thus necessary.

In the plain itself, propagation of the flow Eastwards would be limited by the railway and the motorway embankments which cross the plain from North to South. The 2 dikes are parallel in most part of the plain and separated by a few hundred metres ; the railway is generally lower than the motorway. Consequently, we decide to study only the effect of the motorway embankment on the propagation of the flow. 2 models were thus constituted corresponding to the 2 extreme situations :
- the embankment resist to the flow which cross it by the works (bridges, culverts,..) or by overtopping ;
- the embankment is instantaneously destroyed by the coming flow so that we can consider that the topography is the natural one.

For the first case, the topography of the model included the geometry of the motorway embankment in the simplified way of a trapezoidal shape which becomes 3 cells in the cross direction (1 cell for the top and 1 for each side). In the same way, the mesh of the minor bed was constituted of 7 cells as shown on fig. 2.

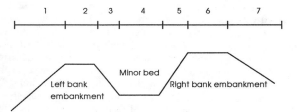

Fig.2-Mesh of river bed

The simplification of the topography which underlies this mesh was justified as the minor bed was widened to trapezoidal cross-sections. In cases where the minor bed is more irregular, it should be necessary to couple 1-D computation of the minor bed to 2-D computation of the surrounding plain. This would be relatively easy to implement as 1-D and 2-D codes lay on similar explicit schemes.

One difficulty of the computations carried out was the overtopping of embankments. As the flow generally went Eastwards, which corresponded to the direction of the minor bed, higher velocities occurred generally over the motorway embankment. They were the cause for the minimum time step observed (below 0.5 s). As we have an explicit scheme, the CFL condition implies $\Delta t < 0.5 \dfrac{\delta}{|v| + c}$ where δ, size of the cell, is about 10 m (instead of 200 m in the inundation plain) and v, average velocity, can reach 10 m/s. The choice of a suitable mesh (with possible topographical simplifications) is thus absolutely necessary to avoid too small time steps.

Fig. 3 illustrates the comparison of the results of the 2 computations with or without the motorway embankment. It is obtained by interpolating the difference between the water level of the first computation (with embankment) and the water level of the second computation on a regular mesh 100 by 100.

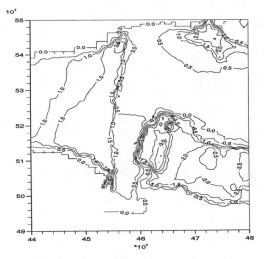

Fig.3-Agly River Dam-Impact of motorway

The motorway embankment cuts fig. 3 from North to South by half whereas Agly River cuts fig. 3 from West to East by half. In the left part (upstream from the motorway), water level is obviously higher if the embankment resist to the flow. In the upper right corner (downstream, left bank), water depths are more important also whereas, in the lower right corner (downstream, right bank), water level is much lower. In fact, as the embankment in the southern part is higher than in the northern part, in case the dike resist, a bigger part of the total discharge is sent to left bank, phenomenon which is not totally compensated by the decrease of the total peak discharge.

SEDIMENT TRANSPORT

The 1-D code RUBAR 3 solves the shallow water equations written in the variables (S, Q) :

$$\frac{\partial S}{\partial t} + \frac{\partial Q}{\partial x} = q \tag{7}$$

$$\frac{\partial Q}{\partial t} + \frac{\partial\left(\frac{Q^2}{S} + p\right)}{\partial x} = gS(I - J) + kqV \tag{8}$$

where S is the cross-sectional area, Q the discharge, I the bed slope, J the bed friction slope generally computed from a Strickler's coefficient, q the lateral inflow, V the average velocity, k a coefficient, p the pressure defined by $\int_0^h g(h - y)L(y, x)dy$ where L is the cross-sectional width, B the lateral pressure defined by $\int_0^h g(h - y)\frac{\partial L}{\partial x}dy$.

As, in some cases, particularly, for dam-break wave, near the dam, sediment transport may be important, we add equations in order to model it. The main objective was to localise deposits and try to foresee a possible rise in maximum water level due to these deposits.

Modelling of the phenomenon was simplified as we refer only to the median grain size diameter (D_{50}). Equation (9) was used :

$$(1 - \lambda)\frac{\partial S_s}{\partial t} + \frac{\partial G}{\partial x} = q_s \tag{9}$$

where S_s is the cross-sectional solid area, λ the porosity, G the sediment discharge (m^3/s), q_s the lateral sediment inflow (m^2/s). For our simplified model, G is computed by the Meyer - Peter - Muller's formula under the following form (see end of text for notations) :

$$G = \frac{8L}{g(\rho_s - \rho)\sqrt{\rho}}\left(\rho gJR - 0.047 gD_{50}(\rho_s - \rho)\right)^{3/2} \tag{10}$$

where J bed friction slope may be computed from the Strickler's coefficient K_f linked to the grain size diameter through $K_f = \frac{21}{D_{50}^{1/6}}$.

S_s is defined by some couples width - elevation (usual topographical description of a cross-section). The evolution of S_s should be transformed in a modification of these couples (L, z). We choose to define a width of the active bed $L_a(x)$. A couple (L, z) becomes (L, z + Δz) where $\Delta z = 0$ if L > L_a and Δz = ΔS_s / L_a if L < L_a.

Equations (7), (8) and (9) are coupled as the computation of p, B and J requires to use the couples (L, z). So, 2 methods may be used to solve the system (7) + (8) + (9) in case of dam-break waves :

- the system (7) + (8) is solved independently as the modification of solid section is much slower than the hydraulic changes. At every time step (or less frequently), the solid section is modified using the results of previous hydraulic

computations. In order to ensure conservation of volumes, the variables S and Q are unchanged when modifying S_S. This implies that the values of the secondary variables h and z are not kept.
- the 3 equations are solved together. Even with the simplifications described here above, the coupling is complicated and solving in such a way would have cost long computation time.
For the example of Lawn Lake Dam described here below, we have thus used the first method. Fluxes for equations (7) and (8) at time $t_{n+1/2}$ are computed in the same way as for fixed bed, i.e. considering that there is no coupling. G is computed in a similar way and couples (L, z) are possibly changed. Then, final computation of S and Q (second member) is done taking into account the evolution of S_S between t_n and t_{n+1}.
Using this method, computation takes about 30% more time than a fixed-bed computation.

LAWN LAKE DAM

Lawn Lake dam is located in Colorado (U. S. A.). When it broke on July 15, 1982, its height was about 8 metres and reservoir contained 830,000 m^3. 3 people died and damages totalled \$31,000,000. R. D. Jarrett and J. E. Costa related ([2]) the dam failure and gave such information as estimated peak discharge at dam site and topographical description of valley downstream from the dam.
First, we modelled this failure using the 1-D code RUBAR 3 with fixed bed and secondly using the same code with movable bed (description here above). In both cases the hydrogramme at dam site was obtained by CEMAGREF's code RUPRO which models the erosion of an earthen dam due to piping ; the properties of the embankment material were adjusted to obtain the same peak discharge as mentioned in [2].
Topography was described by some cross-sections between which computational cross-sections were interpolated at a space step of about 20 metres. Due to the occupation (forest) and the meandering of the valley, a Strickler's coefficient of 10 was chosen.
Computation with fixed bed gave maximum water depths (6 observed points for about 10 kilometres of valley) and arrival time for maximum water depths (2 observed points) in good agreement (difference less than 20 %) with the observations except for maximum water depth at the distance of 2,400 metres from the dam at a site where important deposits were formed.
Fig. 4 shows that computation with movable bed improves the results as it creates an important deposit around the given cross-section. Nevertheless, change in maximum water depth (computed to the initial bed elevation) is not totally satisfying ; similar result can be observed at distance of 900 m.

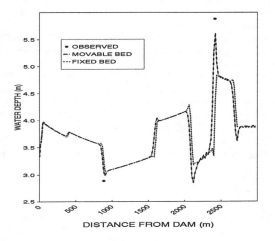

Fig.4-Lawn Lake Dam-Maximum water depths

More generally, the thickness of the layers of deposits (up to 3 m) or the scoured depths proposed by the model agree fairly well with the observations in the described cross-sections. However, they are concentrated in the points where topographical changes occurred in the model which are not actually the only points concerned by deposits or scouring (it was a general phenomenon downstream the dam). A more detailed topography would have certainly improved the results.

In fact, results of the code with movable bed are very much dependant of the following factors :
- choice of D_{50} and of active bed width (here we chose minor bed width) ;
- choice of transport formulas and of their empirical coefficients ;
- form of the hydrogramme at dam-site ;
- local topographical description.

Beyond these aspects which should be solved by the engineer faced to an actual problem , some more theoretical questions remain about the description and modelling of sediment transport, particularly for unsteady flow. In that domain, the two axes of CEMAGREF's research are the evolution of cross-section during a flood and the influence of unsteady flow (accelerations) on the rate of sediment transport.

CONCLUSION

In order to assess risks in case of dam-break, CEMAGREF has developed 1-D and 2-D codes that solve shallow water equations. Extension to the modelling of sediment transport is being implemented. These codes have already been used for more than 20 dams. They could also be applied in any problem of flood when classical implicit schemes fail.

Numerical methods used make computations possible even when hydraulic discontinuities occur. Explicit scheme implies long computations due to the CFL condition. However, it is generally easier to take into account various obstacles such as embankments, bridges,...

The finite volume method makes also possible (without complex algorithm) coupling between 1-D and 2-D computations which is often necessary for modelling flood plains.

REFERENCES

1.   Benoist G., Les études d'ondes de submersion des grands barrages d'EDF, La Houille Blanche, N°1-1989, pp. 43-54.

2.   Jarrett R. D., Costa J. E., Hydrology, Geomorphology, and Dam-Break Modelling of the July 15, 1982 Lawn Lake Dam and Cascade Lake Dam failures, Larimer County, Colorado, U. S. Geological Survey Professional Paper 1369, 1986.

3.   Paquier A., Barrage sur l'Agly : étude de propagation de l'onde de rupture (du barrage à la mer), CEMAGREF groupement de Lyon, 1993.

4.   Van Leer B., Towards the Ultimate Conservative Difference scheme. V. A Second-Order Sequel to Godunov's Method, Journal of Computational Physics, 32, 1979, pp. 101-136.

5.   Vila J. P., Sur la théorie et l'approximation numérique des problèmes hyperboliques non linéaires. Applications aux équations de Saint Venant et à la modélisation des avalanches de neige dense, thèse de doctorat de l'université Paris VI, 1986.

6.   Vila J. P., Modélisation mathématique et simulation numérique d'écoulements à surface libre, La Houille Blanche, N°6/7-1984, pp. 485-489.

7.   Vila J. P., Simplified Godunov schemes for 2x2 systems of conservation laws, SIAM Journal on Numerical Analysis, Vol. 23, N° 6, December 1986, pp. 1173-1192.

NOTATIONS

A_n value of variable A at time t_n
B   lateral pressure
c   celerity
C   Chezy's friction coefficient
D_{50} median grain diameter
g   gravitational acceleration
G   sediment discharge
h   water depth
I   bed slope
J   friction slope
K   Strickler's friction coefficient
L   cross-sectional width
p   pressure
q   lateral discharge
q_s lateral sediment discharge
Q   discharge
R   hydraulic radius
S   cross-sectional area
S_s cross-sectional solid area
t   time coordinate
u   in 2-D, velocity along the x-axis
U   = h u, "discharge" along the x-axis
v   in 2-D, velocity along the y-axis
V   average velocity of cross section ;
       in 2-D, = h v, "discharge" along the y-axis
W_L left value for the variable W in the Riemann problem
W_R right value for the variable W in the Riemann problem
x   longitudinal space coordinate ; distance to dam
y   in 2-D, with "x", space coordinates
z   water-surface elevation
Z   bed elevation
λ porosity
ρ fluid density
ρ_s sediment density
Δt time step
Δx space step
ΔA variation of variable A during Δt

Model Verification against Laboratory and Field Measurements

Computation of a flood event using a two dimensional finite element model and its comparison to field data.

P.D. Bates and M.G. Anderson
Department of Geography
University of Bristol
Bristol BS8 1SS
UK

J.-M. Hervouet
Electricité de France Direction des Etudes et Recherches
78400 Chatou
FRANCE

ABSTRACT

This paper reports on an attempt to begin a rigorous field testing programme for two dimensional finite element flow models, examining the impacts of mesh resolution and numerical method on simulation results. To achieve this a two dimensional finite element model, TELEMAC-2D, has been applied to an 11 km reach of the River Culm, Devon, UK for which field data concerning discharge and flood inundation were available. This data has been used to assess the performance of two new numerical methods, a Streamline Upwind Petrov Galerkin formulation and a hybrid scheme combining characteristics and centred differences, and a special treatment for flood propagation over initially dry domains that have recently become available within the scheme. This has been achieved for two finite element discretizations consisting of 2040 and 9600 elements. The paper shows progress towards a more complete validation of the scheme and demonstrates a critical need to improve field data capture procedures if such progress is to be sustained.

INTRODUCTION

In this paper we address the field testing of two dimensional finite element hydraulic models applied to compound channel flow problems. For this flow problem the standard analytic tool is currently based upon a finite difference solution of the one dimensional St. Venant equation (Samuels[1]). However, two dimensional finite element techniques would appear to have significant potential in this area due to their ability to represent complex topography, such as a river channel meandering within a wider flood plain belt, with a minimum number of elements. The application of two dimensional finite element methods to free surface flows has, until recently, been confined to certain classes of problem. These have include the analysis of detailed flow patterns around structures (eg. Tseng[2]; King and Norton[3]) and river confluence studies (eg. Niemeyer[4]; Su et al.[5]). In all these cases the scale of interest has been small (c. 0.5 - 2 km) with the whole domain fully inundated during a simulation. In such models' representation of the flow field boundary can only be achieved by including or eliminating partially wet elements from the solution domain. The flood plain inundation extent is therefore represented as an irregular front based on the element geometry rather than as a smooth feature. Undesirable, and potentially mathematically unstable, oscillations in the frontal position or spurious flow velocities may occur with such schemes in response to relatively small depth changes. Thus the finite element method as originally developed is of limited use for flood plain studies.

Recent numerical algorithm development, particularly in relation to flow field propagation over initially dry domains, and the availability of enhanced computing power has enabled the application of two dimensional finite element methods to compound channel flow problems at reach scales appropriate to flood inundation phenomena (10 - 30 km). A number of studies have demonstrated that significant potential for this approach exists (Gee et al.[6]; Bates et al.[7]; Feldhaus et al.[8]; Bates and Anderson[9]) and have made limited comparisons to field data. Having demonstrated the potential of two dimensional finite element schemes in this context there is an immediate need to begin a programme of testing to critically assess their utility and contribute to their future development. This is the fundamental research need addressed in this paper.

Accordingly, the two dimensional finite element model TELEMAC-2D has been applied to an 11 km reach of the River Culm, Devon, UK. This has enabled, for the first time, testing of two new numerical solvers and a special treatment for flood propagation and recession over initially dry areas that have recently become available within the code. Computations have been developed for two finite element discretizations of 2040 and 9600 elements constructed for the above reach. The resultant flow field predictions from these models have been compared to data concerning discharge at the downstream outlet of the model and maximum flood inundation extent for an actual 1 in 1 year recurrence interval flow event. This study is then used to make further progress towards the validation and continued development of two dimensional finite element schemes for this class of flow modelling problem.

MODEL DESCRIPTION

TELEMAC-2D solves an equation for two dimensional mass continuity and the two dimensional St. Venant equations for shallow water flows, derived from the full three dimensional Navier-Stokes equations. These are given in conservative form as:

$$\frac{\partial h}{\partial t} + \frac{\partial Q_x}{\partial x} + \frac{\partial Q_y}{\partial y} = 0 \tag{1}$$

$$\frac{\partial Q_x}{\partial t} + \frac{\partial}{\partial x}\left(uQ_x + g\frac{z^2}{2}\right) + \frac{\partial}{\partial y}\left(vQ_x\right) = D_x + S_x \tag{2}$$

$$\frac{\partial Q_y}{\partial t} + \frac{\partial}{\partial x}\left(uQ_y\right) + \frac{\partial}{\partial y}\left(vQ_y + g\frac{z^2}{2}\right) = D_y + S_y \tag{3}$$

Where: u, v = velocity components in the x and y Cartesian directions; h = depth of flow; z = free surface elevation; Q_x, Q_y = discharge components where $Q_x = hu$ and $Q_y = hv$; D_x, D_y, S_x, S_y = diffusion and source terms; g = gravitational acceleration; t = time.

The model therefore solves for the three unknowns h, u and v. TELEMAC-2D also has the possibility to replace h in these equations with the wave celerity, C. This facility gives interesting properties of symmetry and conditioning of the linear systems obtained after discretization but has not been used in the present study.

A mean flow concept is employed to treat turbulent flows, averaging instantaneous velocities and depths over time to give mean motion only. In this formulation an additional term, the Reynolds stress, is added to the governing equations to represent the increased internal shear stress on mean flow produced by velocity fluctuations. As evaluation of this term poses a number of problems for field studies some model of the turbulence must be introduced to render the governing equations mathematically tractable. In TELEMAC-2D the Reynolds stress is assumed to be the product of the depth-mean velocity and an exchange coefficient, ε, dimensionally similar to the coefficient of viscosity, μ, and termed the eddy viscosity. This eddy viscosity term may then be estimated assuming a constant value or calculated from a further equation set such as the k-ε model. For all simulations reported in this study a constant eddy viscosity was employed.

To solve the governing equations TELEMAC-2D (version 2.3) employs a fractional step method where advection terms are solved initially, separate from propagation, diffusion and source terms which are solved together in a second step. This is achieved for a space discretization consisting or linear triangular or linear quadrilateral elements with three of four nodes per element respectively. Several schemes may be used for the advection step, with the characteristics method chosen here for the momentum equation. To ensure mass conservation two alternative schemes are available for the advection of h in the continuity equation; the Streamline Upwind Petrov Galerkin (SUPG) method (Brookes and Hughes[10]) and a hybrid numerical method specifically developed for TELEMAC-2D. According to the SUPG technique, standard Galerkin weighting functions are modified by adding a streamline upwind perturbation. The hybrid numerical scheme consists of a combination of characteristics and centred differences which sacrifices the unconditional stability of the characteristics method for an improvement in mass conservation properties. The second step (propagation) of TELEMAC-2D makes use of a semi-implicit time discretization and solves the resulting linear system with a conjugate gradient type method. In addition, the TELEMAC-2D code makes significant savings in both computational time and storage requirements through the use of an element-by-element solution technique. Here the solution matrices are stored in their elementary form without full assemblage.

TELEMAC-2D incorporates an additional modification to simulate areas of low lateral bed slope such as flood plains or tidal flats. If this option is not implemented for flood plain areas, partially wet elements are retained within the solution domain and the model interpolates a spurious non-zero lateral free water surface across the element. For such an element the remaining terms in the momentum equation are:

$$\frac{\partial u}{\partial t} = -g \frac{\partial z}{\partial x} \tag{4}$$

Where: z = free surface elevation.

Which therefore causes the numerical scheme to overpredict flow velocities. The object of the modification for flood plain areas in TELEMAC-2D is to provide a better approximation of the water surface slope. First, a check is made for each element to test for the existence of a partially wet element. If the bottom elevation for a particular node, z_2 with depth h_2, is greater than the water surface elevation at one of the other nodes in that element with depth h_1, then a new water surface elevation is defined as:

$$z_2 = z_1 + h_2 \tag{5}$$

This can then be used to develop more realistic flow velocities in flood plain areas. Moreover, as only the momentum equation is modified, mass conservation properties are not affected.

STUDY REACH AND MODEL CONSTRUCTION

In order to explore the issues outlined in the introduction to this paper an application of the TELEMAC-2D model was made to an 11 km reach of the River Culm, Devon, UK. This reach consists of a main channel approximately 10 m wide, meandering within a wider flood plain belt up to 450 m in width. This reach was selected due to the regularity with which overbank flooding occurs, with substantial flood plain inundation typically occurring on about 6 occasions per year, and the complexity of the flood plain topography which includes mill races, flood plain embankments and a channel bifurcation. Two finite element discretizations, consisting of 2040 (see Figure 1) and 9600 elements have been constructed for this reach with bottom elevations interpolated using a single topographic representation digitised from 1:2500 series maps and a limited field survey. Meshing of the domain was constrained by the necessity to retain a sufficient density of elements in the main channel area to adequately represent the channel cross section. In the case of the low resolution model this led to the use

of large (150 m by 150 m) flood plain elements and a high length to
width ratio for channel elements. A summary of the characteristics of
each mesh is provided in Table 1.

Mesh resolution	Number of nodes	Number of elements	Element length to breadth ratio	Maximum Courant number (2s time step)
Low	1200	2040	20:1	0.6
High	5600	9600	5:1	0.2

Table 1: Characteristics of the TELEMAC-2D finite element
discretizations constructed for the River Culm, Devon, UK.

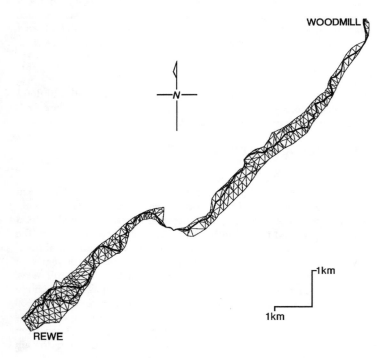

Figure 2: Low resolution finite element mesh constructed for an 11 km
reach of the River Culm, Devon, UK.

Boundary condition data detailing inflows and outflows to the finite element domain were provided by continuous stage recorders at the up- (Woodmill on Figure 1) and downstream (Rewe on Figure 1) extremities of the reach. Rated sections also exist at these sites allowing discharge to be estimated from the stage data. All model simulations reported in this paper were developed using a prescribed flow rate at the upstream boundary and a prescribed water surface elevation for the downstream condition. Discharge data from the downstream gauge were used to independently validate the model. Uniquely, information concerning flood inundation extent during overbank flow events was also available for this reach and was therefore used to provide an additional validation data source.

For each mesh/numerical method combination an initial near steady state flow condition was developed representing flow in the main channel only. This was then used as the starting point for each dynamic flood computation. A boundary friction parameterization was specified using the Strickler coefficient, differentiating between main channel and flood plain regions. Calibration of the model was achieved via trial and error manipulation of the flood plain friction parameter until a satisfactory fit to a test data set was obtained. This process resulted in the selection of Strickler coefficients of 40 for the main channel and 12 for the flood plain.

NUMERICAL RESULTS

Using the above model structure a 1 in 1 year recurrence interval event which took place on 30/1/1990 was simulated for each mesh/numerical method combination. This computation was discretized into 27 000 time steps of 2 s duration, a configuration which appeared to be the optimum for minimising computational time. Results for these simulations are given in Figures 2 and 3 for model predictions of outlet discharge and flood inundation extent respectively. In addition, an example of the ability of the model to simulate the propagation of a flood wave over an initially dry domain is given in Figure 4. Simulations using the hybrid numerical scheme over the low resolution finite element mesh were unsuccessful. A number of problems were apparent for this case including poor mass conservation properties, irregular velocity vectors and significant negative depths. The method was therefore rejected as an appropriate numerical technique for use with such highly distorted meshings. Computation times on a HP9000/735 workstation for the remaining simulations are given in Table 2.

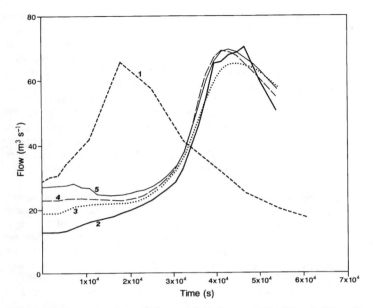

Figure 2: A comparison of observed discharge at the Woodmill and
Rewe gauging stations with TELEMAC-2D predictions where: 1 =
observed upstream discharge at Woodmill; 2 = observed downstream
discharge at Rewe; 3 = predicted discharge using the SUPG method
and a low resolution mesh; 4 = predicted discharge using the SUPG
method and a high resolution mesh; 5 = predicted discharge using a
hybrid numerical scheme and a high resolution mesh.

Mesh/solver	Number of elements	Computation time	Efficiency (Time per 1000 nodes per time step)
Low resolution/SUPG	2040	320 minutes	0.349 s
High resolution/SUPG	9600	2118 minutes	0.480 s
High resolution/hybrid	9600	2930 minutes	0.664 s

Table 2: Computation times and efficiencies for TELEMAC-2D
simulations of a 1 in 1 year recurrence interval flood event discretized
into 27 000 time steps of 2 s duration conducted on a HP9000/735
workstation using a variety of numerical techniques and mesh
resolutions.

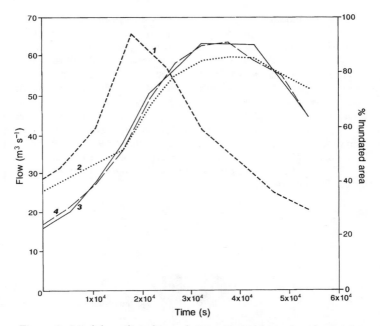

Figure 3: Model predicted inundation extent where: 1 = observed upstream discharge at Woodmill; 2 = predicted inundation extent using the SUPG method and a low resolution mesh; 3 = predicted inundation extent using the SUPG method and a high resolution mesh; 4 = predicted inundation extent using a hybrid numerical method and a high resolution mesh.

Initial evaluation of the flood propagation

Figures 3 and 4 demonstrate that the new algorithm included in TELEMAC-2D to simulate flood propagation over initially dry domains results in simulations of dynamic flood plain inundation extent that are numerically stable and physically realistic. The flood plain inundation front is represented as a continuous feature which advances and recedes laterally in a smooth fashion as the flood wave propagates downstream. Maximum observed flood plain inundation for this particular flood event has been estimated from an amalgamation of data sources, including ground and air photos and post event mapping of flood plain deposits, as being approximately 63 % of total flood plain area. This compares to a model predicted peak of between 80 and 90% depending on the particular mesh/numerical method combination employed. We have therefore been able, for the first time, to compare model predictions of flood propagation to field data. These are shown to be of the correct general order, although a degree of overprediction is evident. It is however difficult to ascertain whether this overprediction is caused by the flood propagation algorithm itself or an inadequate specification of the topographic data used to describe the flood plain surface. Predicted inundation extent, unlike bulk flows, will be highly sensitive to errors in the topographic survey, with variations of a few centimetres potentially causing significant changes in the position of the predicted inundation boundary. Without a fuller understanding of such effects we may merely state that flood propagation as predicted by the model is approximately correct.

(a)

(b)

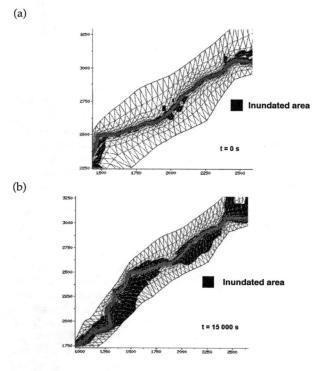

Figure 4: Flood plain inundation predicted by TELEMAC-2D at (a) t = 0s and (b) t = 15 000 s of a 1 in 1 year recurrence interval flood.

Comparison of numerical methods

Figures 2 and 3 show little difference in simulated discharge and inundation extent between calculations made with the SUPG and hybrid numerical solution scheme. In terms of discharge predictions, the SUPG method appears to allow a more convincing initial steady state to be obtained. However, once the rising limb of the hydrograph is reached the two numerical techniques produce near identical results. In terms of computation time, the SUPG method (see Table 2) gives an approximate 28% reduction in computational requirements for this flow problem over the alternative hybrid scheme.

Comparison of finite element meshing strategies

Moving from a low to a high resolution finite element mesh gave relatively little improvement in the accuracy of discharge predictions. The high resolution mesh gives a better fit to the peak discharge, however timing of the peak and the initial part of the simulation are replicated more closely by the low resolution mesh. This probably reflects a need to further refine the calibration for the high resolution mesh. Both finite element meshes do, however, produce good results well within the error limits of the data set. In terms of inundation extent, the high resolution mesh predicts a greater range than the low resolution mesh, albeit with similar timing, and is also able to simulate significantly more exposed flood plain states. This is largely to be expected given the reduced element size.

These results indicate that a good solution of the controlling equations is possible even with a rather coarse discretization of the computational domain and that, in terms of computational cost, these calculations can be achieved relatively cheaply. It is likely, however, that a lack of equivalent high resolution topographic data set is constraining the performance of the high resolution mesh. In this study, therefore, the high resolution mesh is over specified relative to topography. If higher resolution topographic data were available it is probable that the high resolution mesh would give improved performance, particularly in relation to spatially distributed predictions of hydraulic variables.

DISCUSSION

This paper has reported initial efforts to begin a rigorous testing programme for two dimensional finite element models applied to compound channel flow problems. Discharge predictions from the TELEMAC-2D model for a 1 in 1 year recurrence interval flow event have been examined for a variety of mesh resolution/numerical method combinations and found to show a good correspondence to field data. Predictions of inundation extent have also been made using a new approach to flood propagation and recession over initially dry areas that has recently become available within the scheme. Investigation of these predictions has been undertaken and the new algorithm would appear to work well at this particular model scale. Model validation is, however, an ongoing process and much work remains to be done. Related to this validation exercise is the recognition that in this study we have only been able to examine model predictions from a small percentage of the 27 000 time steps calculated during each simulation. This analysis is obviously inadequate if we wish to undertake a complete testing of the model in question. The ability to perform such high temporal resolution calculations thus requires the development of data visualisation methods, such as animation, that allow a fuller understanding of model behaviour.

A need has therefore been identified for further validation and testing of two dimensional hydraulic schemes at spatial and temporal resolutions commensurate with the scale of model simulations. This process should include comparisons between numerical methods and specific algorithms available within particular models, as well as comparisons between different modelling systems. Such research should also make use of a number of test reaches and investigate the impact of domain meshing strategy in a structured way. Data availability is, however, a major constraint to the progress of such research. The River Culm reach represents a site where an atypically high quality data set exists. Despite this, data availability has been shown to be a significant factor in structuring model validation studies and in influencing model development. Resolution of data availability issues is therefore central to the continued refinement of current generation two dimensional hydraulic codes.

NOTATION

C	Wave celerity
D_x, D_y	Diffusion terms in the x and y Cartesian directions
Q	Flow discharge
Q_x, Q_y	Discharge components in the x and y Cartesian directions
S_x, S_y	Source terms in the x and y Cartesian directions
g	Acceleration due to gravity
h	Depth of flow
t	Time

x, y	Cartesian co-ordinates
z	Free surface elevation
ε	Eddy viscosity exchange coefficient
μ	Coefficient of viscosity

ACKNOWLEDGEMENTS

The authors wish to thank Professor Des Walling at Exeter University, UK for making data available to this study. This research was made possible by the support of Electricité de France, Direction des Etudes et Recherches, Paris, France and the Natural Environment Research Council in the UK (Grant Number GR3/8633).

REFERENCES

1.   Samuels, P.G. Cross section location in one dimensional models. In White W.R. (ed), *International Conference on River Flood Hydraulics*, John Wiley and Sons, Chichester, 339-350, 1990.

2.   Tseng, M.T. *Evaluation of flood risk factors in the design of highwater stream crossings. Finite element model for bridge backwater computation*. Office of Research and Development, Federal Highway Administration, Washington D.C., Report No. FHWA-RD-75-53 Vol. III, 1975.

3.   King, I.P. and Norton, W.R. Recent applications of RMA's finite element models for two dimensional hydrodynamics and water quality. *Proc. Second Int. Conf. on Finite Elements in Water Resources*, Pentech Press, London, 81-99, 1978.

4.   Niemeyer, G. Efficient simulation of non-linear steady flow. *J. Hyd. Div. Am. Soc. Civ. Engrs.*, **105**, 185-196, 1979.

5.   Su, T.Y., Wang, S.Y. and Alonso, C.V. Depth-averaged models of river flows. *Proc. Third Int. Conf. on Finite Elements in Water Resources*, University of Mississippi, 1980.

6.   Gee, D.M., Anderson, M.G. and Baird, L. Large scale flood plain modelling. *Earth Surface Processes and Landforms*, **15**, 513-523, 1990.

7.   Bates, P.D., Anderson, M.G., Baird, L., Walling, D.E. and Simm, D. Modelling flood plain flows using a two dimensional finite

element model. *Earth Surface Processes and Landforms*, **17**, 575-588, 1992.

8.   Felhaus, R., Hottges, J., Brockhaus, T. and Rouve, G.   Finite element simulation of flow and pollution transport applied to a part of the River Rhine.   In Falconer, R.A., Shiono, K. and Matthews, R.G.S. (eds), *Hydraulic and Environmental Modelling: Estuarine and River Waters*, Ashgate Publishing, Aldershot, 323-334, 1992.

9.   Bates, P.D. and Anderson, M.G.   A two dimensional finite element model for river flood inundation.   *Proceedings of the Royal Society of London*, **440**, 481-491, 1993.

10.   Brooks, A.N. and Hughes, T.J.R.   Streamline Upwind/Petrov Galerkin formulations for convection dominated flows with particular emphasis on the incompressible Navier-Stokes equations.   *Computer Methods in Applied Mechanics and Engineering*, **32**, 199-259, 1982.

CALCULATION OF 2D FLOOD PROPAGATION USING THE PROGRAM PACKAGE FLOODSIM

Bechteler W., Nujic M., Otto A.J.
Universität der Bundeswehr München
Werner-Heisenberg-Weg 39
D-85557 Neubiberg, GERMANY

ABSTRACT

For flood simulations topographic data and the presence of structures play an important role. Discretising the whole domain two-dimensionally mostly results in an undesireable increase of grid cells and computational time. Even if a very fine grid resolution for the representation of particular structures is used, we are not sure if they are physically well reproduced. Therefore, a more efficient and convenient way is to treat some parts one-dimensionally and to embed these into the 2D model. It turned out that the coupling of 1D and 2D models can be done very efficiently and accurately by using the Poleni formula, especially for rivers or channels the dimensions of which are significantly smaller than the grid size of the 2D grid.

To show the robustness and validity of this approach, a 2D numerical simulation of flooding of a polder beneath the Rhine river has been performed, and numerical results are compared with data from field measurements. Some aspects of 1D - 2D coupling as well as conclusions are discussed.

INTRODUCTION

According to the German IDNDR-Committee for the International Dekade of Catastrophy Prevention, floodings causes the biggest damages all over the world. Therefore, the prediction of the effects of flooding is becoming more and more important for design of hydraulic structures and protection of urban areas. Thanks to the extremly fast development in computer science and remarkable improvements in numerical models, now a powerfull tool for the analysis of flooding problems exists.

At the Institute for Hydroscience Armed Forces University in Munich the program package FLOODSIM has been developed to assist decision making processes with this kind of problems. The program performs a two-dimensional numerical simulation of flooding in complex topographies. It has been already applied successfully to several real case studies, Bechteler et. al. [1,2,3,4] showing its robustness and flexibility.

MATHEMATICAL MODEL

Shallow water flows can be described mathematically by various sets of partial differential equations. A very good review of some possible formulations is given by Weiyan [5]. It is well-known, that the conservation form is preferred to the nonconservation one if strong changes (or discontinuities) in a solution are to be expected, or if mixed flow-regimes are present inside the calculational domain. In flooding problems, this is usually the case because of topographical reasons. Therefore, the following conservation form of the shallow water equations, with (h, uh, vh) as independent variables, has been adopted:

$$U_t + F_x + G_y + S = 0 \tag{1}$$

where

$$U = \begin{bmatrix} h \\ uh \\ vh \end{bmatrix} \qquad F = \begin{bmatrix} uh \\ u^2h+0.5gh^2 \\ uvh \end{bmatrix}$$

$$S = \begin{bmatrix} 0 \\ gh(S_{fx}-S_{bx}) \\ gh(S_{fy}-S_{by}) \end{bmatrix} \qquad G = \begin{bmatrix} vh \\ uvh \\ v^2h+0.5gh^2 \end{bmatrix} \tag{2}$$

Here h represents the water depth, u the velocity component in x direction, v velocity component in y direction and g the gravity acceleration. Bottom slopes in x and y directions are defined as

$$S_{bx} = -\frac{\partial z}{\partial x} \qquad S_{by} = -\frac{\partial z}{\partial y} \tag{3}$$

where z represents the bottom height.
For friction terms Manning's formula is applied

$$S_{fx} = \frac{n^2 \, u \, \sqrt{u^2 + v^2}}{h^{4/3}} \qquad S_{fy} = \frac{n^2 \, v \, \sqrt{u^2 + v^2}}{h^{4/3}} \tag{4}$$

in which n = Manning's roughness coefficient.

NUMERICAL MODEL

The numerical model of the flood propagation (over initially dry area) is based on the high-resolution finite volume method described by Nujic [6]. The numerical scheme uses the Lax-Friedrichs approximate Riemann solver as a building block. We prefer the Lax-Friedrichs solver to the other approximate Riemann solvers (based on flux-difference splitting of Roe [7], or flux-vector splitting of van Leer [8]), not only because of its simplicity or to avoid field-by-field decomposition, but also because it makes it possible to satisfy certain compatibility requirements.

The high resolution of the scheme is achieved by using a kind of the ENO approximation for the numerical fluxes. In the case of a linear hyperbolic equation this scheme is equivalent to the MUSCL scheme of van Leer [9]. The resulting numerical scheme is second order accurate in space and time. The second order accuracy in time is obtained by using a two-stage explicit Runge-Kutta scheme. This time-stepping scheme belongs to the family of Runge-Kutta methods designed by Shu and Osher [10], which preserve TVD properties.

In the reference [6], certain compatibility conditions (or consistency relations) are defined, which should be fullfilled by any numerical scheme in order to get reasonable solutions even on a coarse grid. The importance of this is demonstrated in the same reference. If these conditions are violated, very poor results are usually obtained. In that case, one is forced to use a very fine grid spacing in the neighbourhood of strong changes in the bottom-slope. This increases however the costs of calculation remarkably.

Friction term.
The friction term may cause stability problems as already mentioned by Bechteler et. al. [11]. The reason for this is that when the flow depth becomes very small, the friction term becomes very big. As this term acts in the opposite direction to the flow direction, and because of the discrete time, it may happen that the velocity becomes very big and in the opposite direction to that of the previous time step. Of course this is not a physical phenomena but a pure numerical one, which is simply cured by not allowing friction to change the sign of the velocity during the time step. This is equivalent to the statement that the friction can not take more kinetic energy from the flow as there actually is. Some authors use front tracking and explicitly impose boundary conditions at the front, while Katopodes uses different interpolation procedures in the direction normal to the front and along the front [12]. However, in 2D cases, when the structure of the front becomes very complicated, it is rather difficult to employ these ideas.
Excellent results were obtained using the procedure outlined above, in the

numerical simulation of the dam-break problem described in [11]. The predicted water front as well as water depths fit very good to the experimental results [6].

Initially dry area, drying and flooding and the water film.
The numerical method previousely described preserves positivity of the water depth, if h is used as independent variable in the continuity equation. In the case of variable bottom topography, it is recommended to use water elevation $H=h+z$ as independent variable in the continuity equation, for reasons given in [6]. In this case, positivity of the water depth can not be guaranteed any more. Standard remedy for this is to set a thin film of water over the region. The order of magnitude of this water film varies from 0.1 mm (practically dry area) on a fine grid $dx=dy=2$m, up to 1cm on relatively coarse grid $dx=dy=20$m. Our numerical experiments confirmed that the influence of the water film on the solution is negligible. The numerical experiment has been performed on a very fine grid ($dx=dy=2$m) with constant time step ($dt=0.1$s). The flooding problem considered is described in the following sections. The water film varied between 0.1 mm and 1 cm, and no significant differences between solutions at different time-levels has been found.

COUPLING OF 2D AND 1D MODELS

A flooded area usually contains rivers, channels, roads etc. that may have an essential influence on the process of flooding. All topological features have to be modeled properly to represent the domain in a realistic way. The terrain is mainly calculated two-dimensionally, some parts of it, however, can be described one-dimensionally in a more effective way. Mostly, the flow in rivers can be determined one-dimensionally. In such cases the coupling of the 1D and 2D approaches can save a lot of computational time because the grid size just has to take into account the overall topographical situation of the terrain and is not restricted by the dimension size of small rivers or channels. An application to a flooding problem caused by overflow of the river banks is described in [3,4]. As the dimensions of the flooded area were much larger than the width of river, it turned out to be advantageous to represent the river one dimensional and to embed it into the two-dimensional area. The numerical results were close both to the experimental one and to the result of a truely 2D simulation which was performed on a fine grid, as reported in [3,4].
Fig. 1 shows the treatment of a river, embedded into the two-dimensional computational grid. The river is placed on the cell-faces of the 2D grid, so that the process of coupling is like an internal boundary condition for the 2D model and means lateral inflow or outflow for the 1D model. The quantity of mass and momentum transfer between both models is determined by using the weir formulae for free and drowned flow (Poleni). Other structures inside the domain can be treated in an analogous way.

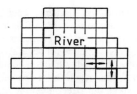

Fig. 1 - 2D Domain with Embedded River

In some cases the existing structure is too singular to be sucessfully treated two dimensionally. In this case the 1D-2D coupling is essential for the simulation, as will be demonstrated by the following examples.

The first example to be considered is a 1D flow over a street. Figure 2a) shows the result of the numerical simulation at q=0.2 m²/s calculated on a very fine grid, dx=0.20 m.

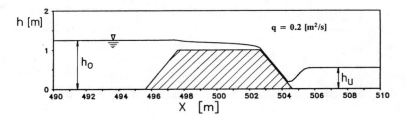

Fig. 2a - Simulated water-level for the flow over a street.

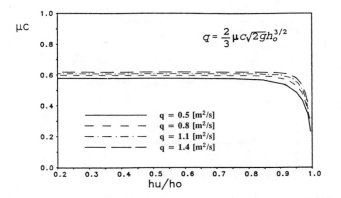

Fig. 2b - Calculated discharge coefficients for the flow over a street for different discharges. h_u/h_o is the ratio of downstream and upstream waterdepth.

As we were interested in calculating the discharge coefficient in the weir formula (see Fig. 2b), it was necessary to use such a fine resolution. Fig. 2b) shows the coefficients, obtained numerically, as a function of the water depth in front of and behind the street. This was done for several discharges and one can see that reasonable values were obtained. Although numerical experiments indicate that it is possible to get similar results on a coarser grid, concerning water depth in front of and behind the street there is still an uncertainty concerning the magnitude of the calculated discharge coefficient.

In the second example a different kind of structure has been chosen as shown on Fig. 3a). A very steep slope at the front side of the weir makes the theory of shallow water equations completely inapplicable for this situation. This can be seen from Fig. 3b), where the numerical discharge coefficient is bigger than one for small discharges. In this case, it is recommended to treat the structure as an internal boundary and to estimate the discharge coefficient to get more realistic results.

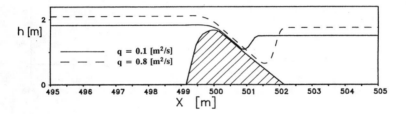

Fig. 3a - Simulated water-levels over a cylindrical-crested structure.

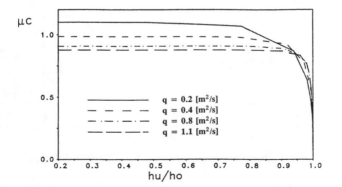

Fig. 3b - Calculated discharge coefficients for the flow over the cylindrical-crested structure in Fig. 3a for different discharges.

Therefore we suggest to treat the structures one dimensional or as internal boundary conditions and, if there is no discharge coefficient available in the case of structures that are not too singular one can try to analyse this coefficient with the help of 1D numerical simulation. In this way a library of discharge coefficients can be built up and used for further applications.

APPLICATION

The numerical scheme, previously described, has been applied for the numerical simulation of flooding for several practical problems. One of these applications will be presented next.

The flooded area "Polder-Altenheim", lies in the south-west of the FRG, besides the Rhein river. The size of the polder is approximately 1x2 km, and its function is to store a certain volume of the discharge from the Rhein during flooding season. A flooding of this area has been performed and measurements were taken at five gauges within the domain. Fig. 4, shows the contour plot of the area and distribution of the Manning coefficient over the domain.

Boundary conditions.

Inlet and outlet fluxes were given (Fig. 6a). Water elevation at inlet/outlet was determined by using zero gradient in the direction of coordinate lines. No other boundary conditions were necessary. This is because the whole area was surrounded by a dam, which has not been overtopped.

The calculation has been performed on a 20x20 m grid what resulted in approximately 4500 grid points. The summer dike, indicated on Fig. 4a, has been treated one dimensionaly as described in the previous section. The small channel on the left side, beneath the Rhine river, has been neglected in the present calculation. This is because, it has almost no influence on the solution at such a big inlet discharge. However, when the ecological flooding of the area is performed, with comparatively small inlet discharges, its importance increases and one should take it into consideration. In this case one can use the embedded approach as described in [3,4].

A quantitative comparison between calculated and measured results (Fig. 5) shows rather good agreement. Especially the time when the measured points (depth indicators) were reached is predicted very well. One possible reason for the small discrepancy between calculated and measured results could be the limited accuracy in determining inlet and outlet fluxes (inlet flux was determined empirically and outlet flux was determined by means of a physical model). The time development of the flooding is shown on the Fig. 6 b. At the first instant of time, water runs along the channel beneath the Rhine and afterwords through the river in the middle of Polder. After five hours the summer dike is overtopped and almost whole area is covered with water.

a)   b)

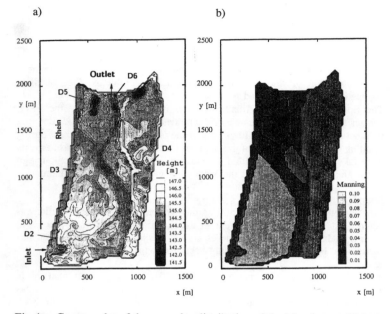

Fig.4a - Contour plot of the area   b - distribution of the Manning coefficient.

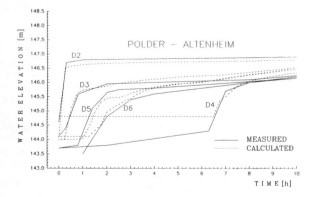

Fig.5 - Calculated and measured water elevations at gauges.

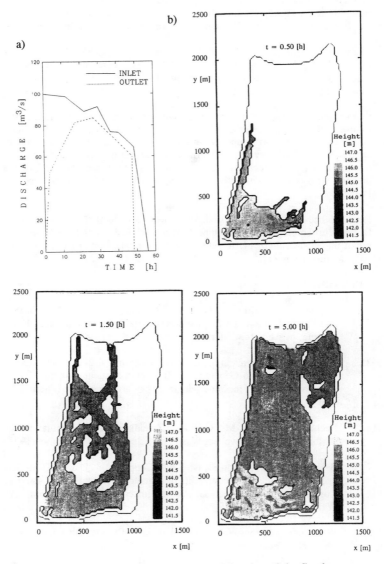

Fig. 6a - Inlet and outlet fluxes,   b - propagation of the flood wave.

CONCLUSIONS

In this article it is shown that a two-dimensional numerical simulation of flooding can be performed very well. Using a reliable and robust algorithm with shock-capturing capability and the ability of handling steep topographical gradients (compatibility relations) is one basic feature of a successful computation. On the other side the modelling of small structures like dams, channels and rivers contributes as well to the quality and effectivness of the simulation.

The coupling approach of 2D and 1D models used in FLOODSIM leads to a reduction of grid cells for the 2D calculation and therefore of the computational time without neglecting the important influence of small structures smaller than the grid spacing.

Storage requirements and the time necessary for preprocessing are minimized by use of the rectangular grid. Nevertheless, it is easy to extend the existing numerical scheme so as to operate on a rectilinear or triangular grid. Such an approach rises flexibility of the algorithm, because it is possible to fit computational grid to the structures inside the domain and to the boundary of the domain. This work is currently under way and the results will be reported later on.

Remark: The authors are very grateful to the "Landesamt für Umweltschutz, Baden-Württemberg - Arbeitsgruppe Hydrologie" and "Amt für Wasserwirtschaft und Bodenschutz - Offenburg" for their cooperation and to provide us with the measuring data.

REFERENCES

1.   Bechteler W., Nujic M., Otto A.J. Program Package "FLOODSIM" and its Application, 1st Int. Conference on Hydro-Science and -Engineering, Washington, edited by Sam S.Y. Wang, 1993.

2.   Bechteler W., Hartmann S., Otto A.J. Coupling of 2D and 1D Models and Integration into Geographic Information Systems, 1st Int. Conference on Hydro-Science and -Engineering, Washington, ed. Sam S.Y. Wang, 1993.

3.   Bechteler W., Nujic M., Otto A.J. Zweidimensionale Simulation der Ausbreitung von Flutwellen, Wasserwirtschaft No. 11, November1993.

4.   Bechteler W., Hartmann S., Otto A.J. Coupling of 2D and 1D Models and Integration into Geographic Information Systems (GIS), 1st Int. Conference on Hydro-Science and -Engineering, York, ed. Sam S.Y. Wang, 1993.

5.   Weiyan, T. Shallow Water Hydrodynamics, Elsevier Sci.Publ. Comp. Inc., New York, 1992.

6.   Nujic, M. Efficient Implementation of Non-Oscillatory Schemes for the Computation of Free-Surface Flows (to be published in Journal of

Hydraulic Research).

7.	Roe, P. L. Approximate Riemann Solvers, Parameter Vectors, and Difference Schemes, J.of Comp. Physics 43, pp. 357-372, 1981.

8.	Van Leer, B. Flux Vector Splitting of the Euler Equations, Lecture Notes in Physics, Vol. 170, pp. 507-512, 1982.

9.	Van Leer, B. Towards the ultimate conservative difference scheme - III, J. Comput. Phys., 23, 1977.

10.	Shu, C.W. and Osher, S. Efficient Implementation of Essentially Non-oscillatory Shock-Capturing Schemes, J. Comput. Phys., 77, pp. 439-471, 1988.

11.	Bechteler, W., Kulisch, H., Nujic, M. 2-D Dam Break Flooding Waves - Comparison between Experimental and Calculated Results, 3rd Int.Conf. on Flood and Flood Management, Florence, 24-26. November, 1992.

12.	Katopodes, N.D. Nearly Authentic 2-D Model for Dambreak Flood Waves, 1st Int. Conference on Hydro-Science and -Engineering, Washington, edited by Sam S.Y. Wang, 1993.

Two-Dimensional Dam-Break Flow Simulation in a Sudden Enlargement

M. Četina and R. Rajar
University of Ljubljana, FAGG
Hajdrihova 28
61000 Ljubljana, SLOVENIA

ABSTRACT

The paper presents the use of a two-dimensional depth averaged mathematical model for an unsteady dam-break flow simulation in a sudden enlargement. Continuity and two momentum equations for depth averaged horizontal velocities u and v are solved numerically by the use of a control-volume formulation of the finite difference approach. The main features of this method are: the space staggered grid with a hybrid scheme, and a fully implicit time scheme. An iterate solution procedure with pressure-correction (SIMPLE algorithm) is used. A basic computer program TEACH was accomplished. It takes into account irregular geometry and time derivatives in the equations; the so called "rigid lid approximation" was also suppressed. Some simple, unsteady, one-dimensional flows in the two prismatic rectangular channels were computed first. The waves simulated by the proposed two-dimensional model were very similar to those predicted by already verified one-dimensional implicit schemes. Finally, an unsteady flow in a rectangular channel with a sudden enlargement was simulated. A qualitative comparison of the results with those observed on a physical model shows good agreement. Qualitatively, the computed maximal water depth in the area was only 3% greater than the experimental value. This is quite acceptable for practical purposes.

INTRODUCTION

Dam-break wave simulation is a very important topic of hydraulic engineering since such computations are prescribed by law in many countries. Usually it is sufficient to use one-dimensional (1D) mathematical models (Rajar [1]) but in some cases two-dimensional

(2D) modelling is required to simulate certain local flow phenomena (Rajar and Četina [2], Četina [3]). One of the problems that has a great practical importance for standard dam-break wave calculations is behaviour of the wave at the transition through a sudden, huge enlargement. Some kilometres downstream of the dams steep and narrow natural valleys often extend into wide and relatively flat flood plains. In such situations, rapid changes in flow velocities, and sudden drop of flow depths have to be taken into account.

The paper presents the use of two-dimensional depth averaged mathematical model to simulate the above mentioned phenomenon. Scientific background of the model with a short description of basic equations, a numerical solution method, initial and boundary conditions are given first. Additionally, the mathematical model was verified by some simple 1D test cases and by the 2D flow observed on the physical model. On the basis of qualitative and quantitative comparisons of the results, the applicability of the model to predict the flow in natural conditions is discussed.

PHYSICAL MODEL

Model Installation

The physical model was constructed at Skopje University, Faculty of Civil Engineering, and detaily described by Popovska [11]. Here only a brief description is given.

The scheme of the model is shown in Fig. 1. It consists of a concrete rectangular, 20m long channel. Water inflows through the pipe ①, and the perforated wall ② into the stabilising reservoir. Then it spills over the Thompson's weir ③ into the main reservoir ④, 1.2m wide and 0.6m deep. From the reservoir exit, closed by the gate ⑤, at 4m the channel ⑥ is 0.4m wide with a bottom slope of 0.2%. Then the channel becomes suddenly enlarged to 2.8m preserving the same bottom slope (⑦). Initially the channel bottom downstream of the gate was dry. The calibrated value of the Manning's roughness coefficient was $n=0.0137$ $sm^{-\frac{1}{3}}$.

The unsteady flow after instantaneous lifting of the gate ⑤ is characterised by a fast change of hydraulic parameters. These imposed a need for measuring instruments of high dynamic sensitivity. Water levels, basic measuring values, were permanently recorded with capacity gauges in 31 points in the first 5.15m of the enlarged channel (see gauges S0 to S31 in Fig. 6). This area was also covered by the mathematical model (⑧). At some points the flow velocities were also measured, but the accuracy was not satisfactory (Popovska [11]).

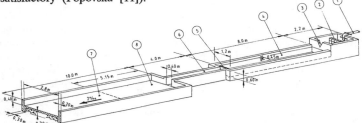

Fig. 1: Scheme of the Physical Model

Description of the Flow Phenomenon

The dam-break waves at three different initial water levels in the reservoir, H_0 =0.45m, 0.35m and 0.25m, were investigated. After instantaneously lifting the gate a positive wave appears downstream. In the enlarged part of the channel it propagates fanlike. From hydraulic point of view the flow in this area is very complicated: it has rapid changes of water levels and those of flow velocity, particularly at the beginning of the simulation. The whole phenomenon on the physical model lasted 50-200s depending upon the initial conditions in the reservoir. For flow analysis purposes, the first 5-10s of the simulation are the most interesting. Namely, the time required for the simulated wave to reach downstream profile of the measuring plane is 4.5s for H_0 =0.45m, 5s for H_0 =0.35m and 6.7s for H_0 =0.25m. Gradually, the flow calms down and becomes uninteresting from hydrodynamic point of view.

In the narrow part of the channel, the flow is one-dimensional. Depth measurement at the place of sudden enlargement shows retaining of maximum levels from 7 to 10s and then gradual decreasing. Measured velocities at this place show supercritical flow during the entire simulation period.

Some local flow phenomena were also observed. They were manifested as waves reflected from the lateral walls that were superimposed on the main wave and formed some kinds of standing waves.

MATHEMATICAL MODEL

Basic Equations

The following assumptions were taken into account: a. Velocities are depth averaged, b. Pressure is assumed as hydrostatic, c.

Manning's quadratic law is used for bed friction, d. Coriolis acceleration is neglected, e. Wind shear stress at the water surface is neglected, f. Boussinesque approach of the effective turbulent stress is adopted.

With the above mentioned assumptions, the 2D continuity and momentum equations describing unsteady flow can be written in Cartesian coordinate system in the following form:

$$\frac{\partial h}{\partial t} + \frac{\partial (hu)}{\partial x} + \frac{\partial (hv)}{\partial y} = 0 \tag{1}$$

$$\frac{\partial (hu)}{\partial t} + \frac{\partial (hu^2)}{\partial x} + \frac{\partial (huv)}{\partial y} = -gh\frac{\partial h}{\partial x} - gh\frac{\partial z_b}{\partial x} - ghn^2 \frac{u\sqrt{u^2+v^2}}{h^{4/3}} + \frac{\partial}{\partial x}(hv_{ef}\frac{\partial u}{\partial x}) + \frac{\partial}{\partial y}(hv_{ef}\frac{\partial u}{\partial y}) \tag{2}$$

$$\frac{\partial (hv)}{\partial t} + \frac{\partial (huv)}{\partial x} + \frac{\partial (hv^2)}{\partial y} = -gh\frac{\partial h}{\partial y} - gh\frac{\partial z_b}{\partial y} - ghn^2 \frac{v\sqrt{u^2+v^2}}{h^{4/3}} + \frac{\partial}{\partial x}(hv_{ef}\frac{\partial v}{\partial x}) + \frac{\partial}{\partial y}(hv_{ef}\frac{\partial v}{\partial y}) \tag{3}$$

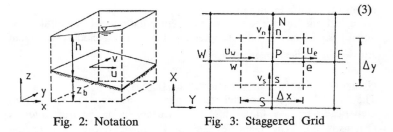

Fig. 2: Notation   Fig. 3: Staggered Grid

where t is the time, h the water depth, u and v the velocity components in the x and y directions, z_b the bottom level, n the Manning's friction coefficient, g the acceleration due to gravity and v_{ef} an effective coefficient of viscosity that will be described in the next chapter.

It is important to notice that due to large energy losses at the transition from supercritical to subcritical flow the dynamic equations (2) and (3) were used in the conservative form.

Turbulence Closure Model

If we neglect laminar viscosity v than the effective viscosity coefficient $v_{ef} = v + v_t$ consists of turbulent viscosity v_t only. The

latest is not the property of the fluid but the property of the flow. Relatively little general remarks about the influence of the terms including ν_{ef} can be found in the literature. They are often neglected, particularly at unsteady, depth-averaged, 2D dam-brake flow (e.g. Popvska [4], [11], Uan [5]).
In some flows including recirculating zones, effective turbulent stresses can play an important role and influence the results significantly (e.g. Rodi and McGuirk [6], Četina and Rajar [7]).
In our case, the small recirculating areas at both edges of the enlarged channel were observed on a physical model. Therefore sensitivity analysis on influence of the value of ν_{ef} was made. We tested the values varying from $\nu_{ef}=0$ to ν_{ef} resulting from the well known depth averaged version of the $k-\varepsilon$ turbulence model (Rodi [8], Četina [3], Četina and Rajar [7]).
The computed velocity fields at three different steady flow discharges were nearly the same either by using the value $\nu_{ef}=0$ or $k-\varepsilon$ turbulence model. The results for the discharge $Q=0.00447$ m^3/s are shown in Fig. 7.
So in all the further unsteady flow computations we simplified the equations (2) and (3) by suppressing the terms including turbulent stresses in vertical planes ($\nu_{ef}=0$).

Numerical Method and Computational Details
The set of coupled partial differential equations 1-3 was solved numerically by the finite difference scheme using control volumes (Patankar [9]). The main features of the method are: a staggered grid (Fig. 3), a hybrid scheme (combination of central difference and upwind schemes) and an iterate procedure based on depth corrections h' (SIMPLE algorithm).
For the integration in time, a fully implicit scheme was used ensuring stable and accurate solutions even at relatively high Courant numbers (up to 15). As the convergence criteria, it was assumed that in every time step the sum of errors in all the points should not exceed 1% of the total discharge through the computational domain.
Due to symmetry only one half of the area was covered by the non-uniform numerical grid (Fig. 6). In the case of steady flow computations, both narrow and enlarged parts of the channel were included in the computation. Thus, the numerical grid had 41 points in the y direction along the flow and 15 points in the cross

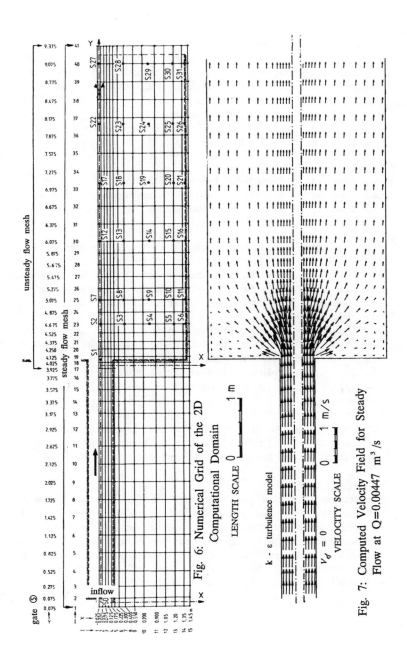

Fig. 6: Numerical Grid of the 2D Computational Domain

Fig. 7: Computed Velocity Field for Steady Flow at $Q=0.00447$ m³/s

direction x. The space step in the y direction varied from $\Delta y = 0.1$m to 0.3m and in x direction from $\Delta x = 0.05$ to 0.2m (Fig. 6). In the case of unsteady flow computations the flow in the enlarged channel was simulated only. The numerical grid was reduced and had 25 points in the y direction ($\Delta y = 0.1$ to 0.3m) but remained unchanged in the cross direction.
The time step $\Delta t = 0.5$s was used. The average values of Courant numbers were among 2 and 3 (but up to 13 at some points).

Boundary and Initial Conditions
Equations 1-3 are hyperbolic, but an elliptic problem is solved at each time step. The values of the depth averaged velocities u, v and depth h are required at all boundaries. In addition, the initial state of the flow must also be known.

Boundary Conditions. Velocities, normal to closed boundaries, were taken as 0. For the longitudinal velocities, simplified conditions of zero normal gradients were assumed. At the inflow boundary the measured hydrographs $h = h(t)$ were prescribed. At the outflow open boundary zero longitudinal gradients of u, v and h were taken into account.
In the cases where $k - \varepsilon$ turbulence model was used, more realistic longitudinal velocities near the wall, resulting from the logarithmic wall function, were assumed (Četina [3]).
At the symmetry plane, the normal velocities and the normal gradients of all the other variables (longitudinal velocities, h, k and ε) were zero.

Initial Conditions. At the beginning of the simulation (time $t = 0$) the distribution of h, u and v has to be known. This initial state is the result of previous steady state computations.

Computer Code
The computer program used was based on TEACH code (Gosman and Ideriah [10]). The possibilities of irregular geometry, depth averaged version of the $k - \varepsilon$ turbulence model, and different boundary conditions were implemented. The basic program was also completed by the possibility of considering significant changes in water surface levels (the so-called "rigid lid" approximation was suppressed) and by adding the unsteady terms.

VERIFICATION OF THE MODEL

1D Test Cases

For the first test, some simple unsteady 1D flows in two prismatic rectangular channels were computed by our 2D mathematical model. Channel 1 was 15m long and 3m wide with a uniform longitudinal bottom slope of 0.36%. For Channel 2, the parameters of the enlarged part of the channel at physical model were taken (length of 5.25m and bottom slope of 2%), while the width of the channel was assumed to be 0.2m. At the inflow boundary the curve $h = h(t)$ registered on gauge S1 was prescribed (Fig. 4). The initially dry bottom was replaced by small initial depth being $h_0 = 0.001$ m.

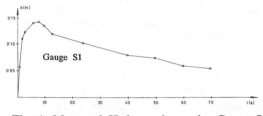

Fig. 4: Measured Hydrograph on the Gauge S1

Results

Several computations with different time steps and water depths

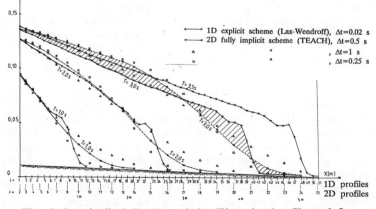

Fig. 5: Longitudinal Profiles of the Wave in the Channel 2

were carried out (Četina [3]). The results were compared with already verified 1D mathematical models using both explicit and implicit schemes. For the Channel 2, the comparison is shown in Fig. 5.

The most important conclusion is that the waves simulated by our 2D model are very similar to those predicted by 1D implicit schemes. In the case of small initial depths, our proposed model is even superior because there are no stability problems.

MODEL RESULTS

As the last step, unsteady flow in the sudden enlargement of the rectangular channel was simulated. Initial depth in the reservoir was 0.45m while the bottom of the channel downstream of the gate was dry. For the computational reasons, a minimal depth of 0.001m instead of dry bottom was assumed.

Steady Flow

Steady flow with the small discharge of $Q=9.63 \cdot 10^{-5} \text{m}^3/\text{s}$ was computed first since it was required as an initial condition for unsteady flow computations. The discharge corresponds to normal depth $h=0.001$m in the wide channel. Figs. 8 and 9 show the results of the steady flow with a considerably larger discharge $Q=0.084 \text{m}^3/\text{s}$, that was 75% of the maximum unsteady flow discharge at the point of the sudden enlargement.

The steady flow computations gave physically realistic results with the appearance of a short area of supercritical flow just after the sudden enlargement. This was confirmed by observations at the physical model and by the detailed discussion of Rajar [1].

Unsteady Flow

Unsteady flow computations were much more time consuming. For that reason, only the enlarged part of the channel was modelled, as shown in Fig. 6. The h-t curve from the gauge S1 (Fig. 4) was considered as the upstream boundary condition. Qualitative and quantitative comparisons of the computed results (some of them are shown in Figs. 10-14) with those observed from the physical model gave some important conclusions that could be summarised as follows:

Qualitatively, the mathematical model is capable to predict the reflection of the wave from lateral walls; the supercritical flow, just after the enlargement with an almost sudden rise of depths at the

- 276 -

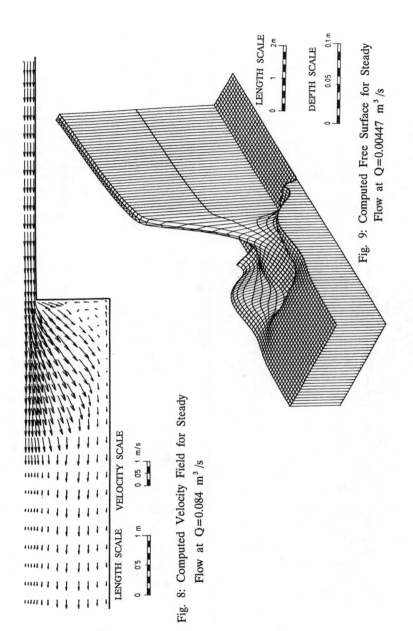

LENGTH SCALE
0   1   2m

DEPTH SCALE
0   0.05   0.1m

Fig. 9: Computed Free Surface for Steady
Flow at Q=0.00447 m^3/s

VELOCITY SCALE
0   0.05   1 m/s

LENGTH SCALE
0   0.5   1 m

Fig. 8: Computed Velocity Field for Steady
Flow at Q=0.084 m^3/s

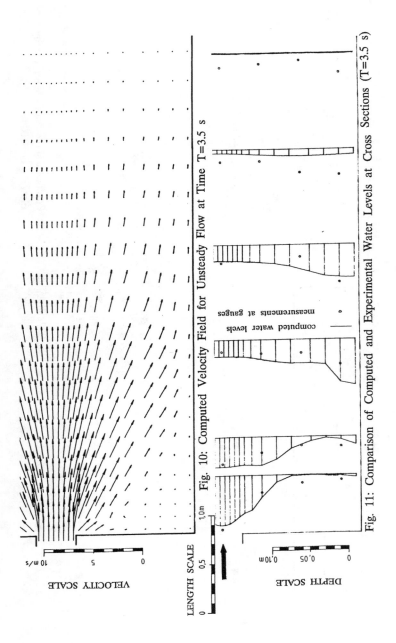

Fig. 10: Computed Velocity Field for Unsteady Flow at Time T=3.5 s

Fig. 11: Comparison of Computed and Experimental Water Levels at Cross Sections (T=3.5 s)

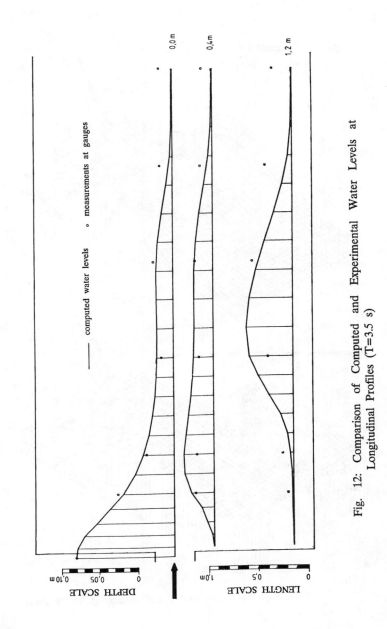

Fig. 12: Comparison of Computed and Experimental Water Levels at Longitudinal Profiles (T=3.5 s)

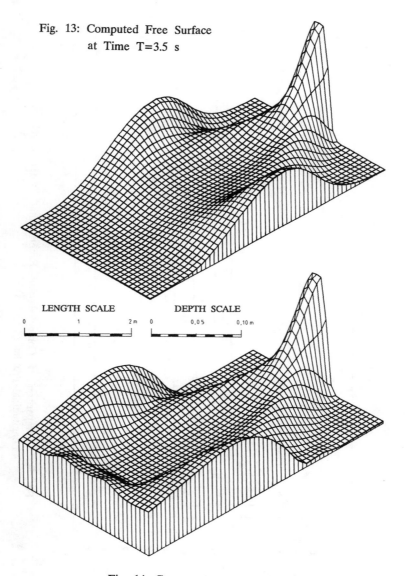

Fig. 13: Computed Free Surface
at Time T=3.5 s

LENGTH SCALE   DEPTH SCALE

0   1   2 m   0   0.05   0.10 m

Fig. 14: Computed Free Surface at Time T=10 s

transition to a subcritical flow; and the small recirculating zones in the corner of the enlargement.

Quantitatively, the computed wave front velocity in the channel axis is underpredicted by about 28%. Genarally, the computed depths are to a certain degree greater than the observed ones. Differences in the maximal computed and observed depths at the gauges on the central axis are within ±30%, and at the gauges near the wall +25%. Hovewer the maximal depth in the area, resulting from the mathematical model, is only 3% greater than the experimental value.

Considering that the maximal depths are of primary importance during the movement of a dam-break wave through a sudden enlargement, our proposed depth-averaged mathematical model gives quite acceptable results for practical use. Of course, further efforts should be undertaken to reduce the differences.

CONCLUSIONS

Applicability of an unsteady 2D depth-averaged mathematical model for simulation of a suddenly enlarged dam-break flow was studied. The analysis of the computational results and their comparison with the selected 1D flows and that of experimental results gave the following conclusions:

1. In comparison with 1D implicit schemes the proposed 2D model is superior because there are no stability problems at small depths. The simulation of supercritical flow is also possible.

2. The transition of the dam-break wave through sudden enlargement is qualitatively well predicted. The quantitative results are acceptable for practical purposes.

ACKNOWLEDGEMENT

The authors wish to express gratitude to Dr. C. Popovska, the University of Skopje, for providing them with the results of her experimental work.

REFERENCES

1. Rajar, R.: *Recherche theorique et experimentale sur la propagation des ondes de rupture de barrage dans une vallee naturelle*, Thesis, Université Paul Sabatier, Toulouse, 1972.

2. Rajar, R. and Četina, M.: *Two-Dimensional Dam-Break Flow in Steep Curved Channels*, 20th IAHR Congress, Moscow, 5.-9. 9. 1983, Proceedings, Vol. II, pp. 571-579.

3. Četina, M.: *Mathematical Modelling of Two-Dimensional Turbulent Flows*, Acta hydrotehnica, No. 5, Laboratory of Hydromechanics of the University of Ljubljana, 1988 (56 pages in Slovene with 5 pages of extended summary in English).

4. Popovska, C.: *Mathematical Model for Two-Dimensional Dam-Break Propagation*, Hydrocomp '89, Dubrovnik, 13.-16. 6. 1989, Proceedings, pp. 193-202.

5. Uan, M.: *Model Bidimensionnel d'Onde de Rupture due Barrage, Comparison Mesures-Calcul Pour un Cas Shematique*, EDF-LNH, E 43/81-08, Chatou, France, 1981.

6. Rodi, W. and McGuirk, J.J.: *A Depth-Averaged Mathematical Model for the Near Field of Side Discharges into Open Channel Flow*, J. of Fluid Mech., 86, 1978, pp. 761-781.

7. Četina, M. and Rajar, R. *Mathematical Simulation of Flow in a Kayak Racing Channel*, Int. Conference on Refined Flow Modelling and Turbulence Measurements, Paris, 7.-10. 9. 1993, Proceedings, pp. 637-644.

8. Rodi, W.: *Turbulence Models and Their Application in Hydraulics*, A State of the Art Review, IAHR Book Publication, Delft.

9. Patankar, S. V.: *Numerical Heat Transfer and Fluid Flow*, McGraw Hill Book Company, 1980.

10. Gosman, A.D. and Ideriah, F.J.K.: *TEACH-T: A general Computer Program for Two-Dimensional Turbulent Recirculating Flows*, Dept. of Mechanical Engineering, Imperial College, London, 1976.

11. Popovska, C.: *Numerical and Experimental Simulation of Two-Dimensional Dam-Break Propagation*, University Kiril and Metodij in Skopje, Faculty of Civil Engineering, Skopje, April 1989.

Formation and Propagation of Steep Waves: an Investigative Experimental Interpretation

P.G. Manciola
Institute of Hydraulics, University of Perugia
Via Borgo XX Giugno, 74
06123 Perugia ITALY

A. Mazzoni
Freeland Engineer
Piazzale Giotto, 24
06100 Perugia ITALY

F. Savi
Institute of Agricultural Hydraulics, University of Milan
Via Celoria, 2
20133 Milan ITALY

ABSTRACT

From the end of the last century, many Authors have
studied experimentally the formation and the
propagation of steep waves on dry flume bottom. Up to
the present time, however, some aspects of this
phenomenon have not been sufficiently explained.
The movement of the water immediately after rapid gate
opening is mainly controlled by vertical acceleration
due to gravity. It follows that the gradually-varied
flow hypothesis is not valid in this initial phase and
the motion is two-dimensional.
Moreover, the type of gate and the velocity of opening
influence the flow across the section and the wave
celerity.
Investigative experiments were conducted in a
rectangular flume with the aim of studying in detail
the flow through the gate section and the celerity of
propagation of the wave immediately after the lifting
of the gate.
The tests were conducted at the Department of
Hydraulic and Enviromental Engineering at the
University of Pavia by removing a top hinged sluice
gate which dams in different sections a flume with
horizontal bottom. A series of tests was carried out

by using an adverse slope base at the downstream reach of the flume. The experiments were observed through the transparent walls of the flume by using three video cameras synchronized with the gate opening device. The video images, digitized directly by the video recorder, were processed and vectorized.

In the gate section, the discharge showed a clear peak immediately after opening and thereafter it stabilized at the critical-state theoretical value.

Other tests were conducted imposing a wave reflecting wall at the downstream section of the flume. The mixing of flows caused by the superimposition of waves travelling in opposite directions was studied by injecting a tracer.

The experimental investigations were interpreted with the aid of a mathematical model which integrates the unsteady motion equations of a gradually varied free surface flow according to a finite difference scheme.

THEORETICAL AND EXPERIMENTAL CONTRIBUTIONS

Many Authors have dealt with the theoretical study of the formation and propagation of steep waves. The results obtained have shown the fundamental characteristcs of this phenomenon.

The movement of water immediately after the lifting of the gate is mainly controlled by vertical acceleration due to gravity. Pohle [12] gives the analytical solution in this initial stage by integrating the two-dimensional irrotational flow equations of an ideal fluid expressed in Lagrangian form.

The vertical acceleration decreases rapidly and when it becomes sufficiently small, de Saint Venant equations can be applied.

Ritter [12] developed an analytical solution of these equations assuming: rectangular section, flat bottom, absence of energy losses. De Marchi [2] generalized Ritter's interpretative scheme for more complex cases.

Dressler [3] obtained an analytical solution which took into consideration the frictional resistive effects. The energy losses were estimated using Chezy's formula.

Whitham [11] divided the wave profile into two regions. In the upper region resistance effects are assumed negligible, so the Ritter's solution can be applied. In the second region, near the head of the wave, momentum and continuity equations in integral form are applied. In the analysis of Dressler and Whitham, channel of infinite width was assumed.

Su and Barnes [9] extended Dressler's solutions in order to include both the effects of resistence and variation of channel cross section.

The first experiments on Ritter's solution were conducted by Schoklitsch [8]. He used a 26 m long flume and a 150 m long by 1 m wide canal. His experiments were particularly directed at verifying the theoretical results near the gate section.

A few years later, Trifonov [10] carried out some similar, but more complete, experiments. He used a flume 30 m long and 0.4 m wide with slope $S_0=0.04$ with a vertically moving sluice gate positioned in the central zone of the flume. He also used two bottom with different roughness. Trifonov found that the average discharge was always lower than the theoretical discharge derived from Ritter's solution. He also observed that the average celerity of the wave front was considerably lower than the theoretical value deduced by Ritter.

Dressler [3] compared his analytical solution with experiments carried out in a horizontal rectangular flume 65 m long and 0.225 wide. Three different flume bottoms varying from smooth to very rough were used.

Faure e Nahas [4] carried out experiments for the propagation of wave front advancing over dry and wet bottom. They used a rectangular flume 40.6 m long and 0.25 m wide and a sluice gate moving upwards. The path of the wave front was in good agreement with the solution computed according to Whitham's theory.

U.S. Corps of Engineers [1] conducted experiments in a rectangular flume 122 m long 1.22 m wide with slope $S_0=0.005$. The gate, located midway of the flume, was lifted upwards. Various test conditions, each representing a different breach pattern, were simulated.

Montuori [5, 6] used a rectangular flume about 26 m long and 0.3 m wide with slope $S_0=0.003$. He produced a steep wave formation by rapidly introducing a constant discharge at the head of the flume. The experiments were conducted with several discharges and with two flume bottoms with different roughness. Montuori's experiments showed that the velocity of the wave front on dry bottom tends to the normal velocity for the constant discharge introduced upstream.

EXPERIMENTAL AND MEASURING APPARATUS

The experimental part of this work was conducted in the laboratory of the Department of Hydraulic and Enviromental Engineering at the University of Pavia.

The formation of a steep wave was simulated by rapidly opening a top hinged sluice gate which dams a rectangular horizontal flume with still water upstream from the gate. The flume is 9 m long and 0.49 m wide. The bottom is made in sheet steel and the side walls are made from 1.2 m wide glass sheets. The side walls of the flume thus consisted of 15 rectangular windows. The gate was installed at various section along the flume. The opening of the sluice gate was effected by a release of a traction spring (UNI 8526 type 6.3x56.3x10.2 cm) applied above the rotation axis.To effect the release of the spring, it was loaded with a force of 1200 N. The device was equipped with a damper which prevented the gate from falling back into the flume after full opening.

The tests were carried out by using three different flume ends: an open end, a closed end (reflecting), a final stretch (3 m long) with an adverse slope. On the flume bottom, tracer was injected in two sections as showed Fig. 1.

The experiments were recorded by three professional quality video cameras positioned on the sides of the flume. The cameras allowed video tape recording at 25 frame/s. Since one frame consists of two images, the time interval of observation was 0.02 s/images.

The video tape was later post-processed in a professional editing studio. The numerical data were extracted both directly from the video and from digitized images. The differences between the two measuring techniques were negligible.

In some images it was difficult to take measurements for the following reasons:
- considerable quantity of foam near the wave front,
- surface tension at the water-glass boundary. This caused a thick bright strip which was visible near the water surface,
- presence of splashes caused by localised irregularities in the flume bottom.

The recording apparatus used, after some initial difficulties, proved extremely effective in both the quality and quantity of images produced.

DESCRIPTION OF EXPERIMENTS

The experiments were carried out with four values of the initial water depth upstream from the gate: $H_0=0.35$ m, $H_0=0.30$ m, $H_0=0.22$ m, $H_0=0.20$ m. The gate was located in two positions: x=3.366 m (pos. A), and x=5.876 m (pos. B). The spatial coordinate x was measured from the upstream section of the flume.
Fig. 1 shows a lateral view of the flume.

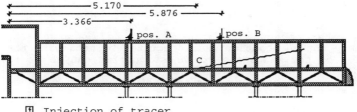

⊞ Injection of tracer

Fig. 1: Lateral view of the flume

Downstream from the gate, both dry bottom conditions and wet bottom conditions were considered. In this case the initial water depth was 0.021 m.

The propagation on adverse slope was carried out by putting a metal slab on the bottom of the flume at x=5.17 m (section C). The slope of 3 m long slab were $S_0=-0.084$, $S_0=-0.096$ $S_0=-0.15$.

The effect of the opening velocity of the gate was also studied. Fast, slow and medium openings were considered.

The experimental conditions are summarized in Table I, which shows the position of the gate, the value of H_0, the conditions of the bottom, the opening velocity, whether reflection took place and whether tracer was injected. The experiments were repeated several times: 40 tests were carried out in all.

Table I: List of the experimental conditions

Gate	H_0 (m)	Bottom	Opening	Refl.	Tracer
A	0.35	dry	fast	no	no
A	0.20	dry	fast	no	no
A	0.35	dry	fast	yes	no
A	0.20	dry	fast	yes	no
A	0.35	wet	fast	yes	yes
A	0.35	dry	fast	yes	yes
A	0.20	dry	fast	yes	yes
B	0.35	dry	fast	no	no
B	0.20	dry	fast	no	no
B	0.35	dry	fast	yes	no
B	0.35	dry	fast	yes	yes
B	0.35	wet	fast	yes	yes
B	0.20	slope	fast	no	no
B	0.30	slope	fast	no	no
A	0.22	dry	fast	no	no
A	0.35	dry	medium	no	no
A	0.35	dry	slow	no	no

In order to estimate the value of the roughness coefficient, some preliminary experiments in steady-state conditions were conducted. For eight values of discharge varying from $7.6 \ 10^{-3}$ m³/s to $61 \ 10^{-3}$ m³/s, the water depths along the flume were measured. The steady motion equation was integrated by modifying the value of the Manning's roughness coefficient so as to fit the experimental data. For the examinated discharge range we estimated the value of n to be equal 0.015 sm$^{-1/3}$.

INTERPRETATION OF THE EXPERIMENTAL RESULTS

In the following paragraphs, the experimental results are illustrated separately, distinguishing among the initial movement of the water, the propagation (either on a horizontal or an adverse slope) and the reflection of the wave at the downstream boundary.
Some experiments were interpreted by means of a mathematical model which integrates the unsteady motion equations of a gradually-varied, one-dimensional flow. These equations are written in conservative form and are expressed in discrete form according to a staggered grid which distinguishes between those sections where water depth is computed from those sections where discharge is computed.
For the features of the scheme and stability analisys see Natale and Savi [7]. It is worthwhile, however, to emphasize that this model allows the simulation of wave propagation on dry bottom and the location of hydraulic jumps. The model is able to describe the propagation of these jumps automatically without imposing internal boundary conditions across the discontinuity.

INITIAL PHASE OF THE MOVEMENT

The discharge at the gate section was estimated in several ways on the basis of the water volumes flowing through the section for various time steps. These volumes were evaluated by measuring the depths of the wave profile downstream from this section.
The instantaneous discharge was evaluated by:
- fitting the experimental volumes to a polynomial function of time (in a few cases 2 polynomials) and computing the first derivative of these equations (Q_{i1}),
- computing the ratio between the variations in time of the measured volumes and the time steps (Q_{i2}).
The average discharge (Q_a) was evaluated by computing the ratio between the measured volumes and the time from the beginning of motion.

Fig. 2. shows the measured discharge hydrographs compared with the value computed according to the analytical solution. The discharge oscillates, showing a clear peak and a less-pronunced trough. These fluctuations are less significant for Q_a but neverthless there is a clear peak.

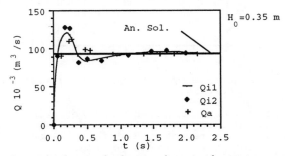

Fig. 2 - Discharge hydrographs at the gate section

This maximum discharge is greater, both for $H_0=0.35$ m and $H_0=0.20$ m, than the analytical value of about 30%. The experimental discharge becomes similar to the analytical solution after approximately 1 s.
With the aim of evaluating the influence of the opening velocity of the gate, experiments were carried out by reducing the tension of the gate spring in order to obtain three different opening velocities. The previous results are obtained with fast opening.
For the different openings, the horizontal velocity w of the lower edge of the gate was measured. For example, after 0.10 s the values of w were 5 m/s, 3 m/s, 1.3 m/s respectively for fast, medium and slow opening. The water is completely detached from the gate respectively after 0.1 s, 0.14 s, 0.20 s. Therefore the openings can be considered sufficiently rapid although not instantaneous.
When the opening velocity was reduced, the behaviour of the discharge does not change (Fig. 3), but the maximum value decreased with a slower opening.
In the following only the results obtained with fast opening are illustrated.

MODELLING OF FLOOD PROPAGATION

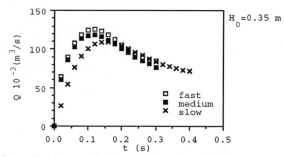

Fig. 3 - Discharge hydrographs for different opening velocities

The mathematical model, although integrating the one-dimensional motion equations, permits the discharge hydrograph to be reconstructed fairly accurately and reproduces the peak of discharge (Fig. 4).

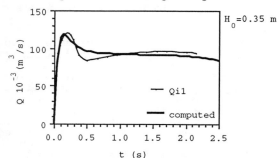

Fig. 4 - Comparison between computed and measured discharge hydrographs

When maximum discharge occurs, the celerity of the wave front propagation shows a pronounced peak. As the wave propagates downstream, the celerity progressively decreases to one third of the maximum value (Fig. 5). The celerity c_1 of the wave front was evaluated by measuring the position of the wave front over time, fitting these values with two polynomials (in order to obtain a satisfactory interpolation both near and far from the gate section), and calculating the first derivative.
Some values of the celerity c_2 were also evaluated by calculating the ratio between variations in the position of the front over time and the time steps.
The variation of celerity was not so accurately reconstructed (Fig. 5), since the model considerably

underestimates the maximum value, which probably can be better simulated with a two-dimensional model.

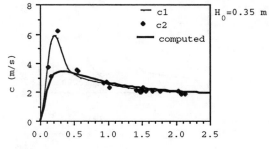

Fig. 5 - Comparison between computed and measured wave front celerities

In the gate section, the water depth h progressively decreases and only after a variable time interval it stabilizes at the critical value given by Ritter. Fig. 6 shows the stage hydrographs computed and measured.

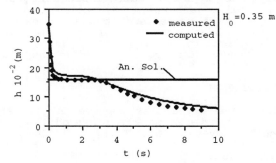

Fig. 6 - Comparison between computed and measured stage hydrographs

It seemed interesting to duplicate Dressler's [3] experiments, which analyzed the path of the sharp corner at the top surface next to the gate.
Taking $H_0=0.22$ m, Fig. 7 shows the path of this point compared with Dressler's results for the same initial water depth (the coordinate x_0 is measured from the gate). Data is only possible up to 0.09 s since the initial shape of the sharp corner tends to change over time.

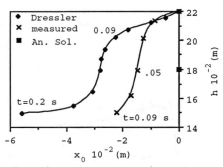

Fig. 7 - Sharp corner trajectory

Our experiments showed a vertical displacement greater than both Dressler's results and the value given by the analytical solution. Dressler explained that the differences between his results and Pohle's solution where caused by both surface tension effects, which are predominant at the sharp corner, and by the small dimensions of the impoundment upstream from the gate. Probably the type of gate influences the path. Dressler used a gate which was pulled up by a system of springs. As a consequence the water particles are in contact with the gate for a longer time than in our case. It therefore follows that the gate used by Dressler tends to support the water particles thus reducing vertical acceleration.

WAVE PROPAGATION

In order to describe this phenomenon accurately, we distinguished between wave routing on the horizontal slope and on adverse slope.

Propagation on horizontal bottom
Immediately after the gate starts opening, air enters between the gate and the bottom of the flume, deforming the shape of wave front. Fig. 8 shows the characteristic cusp shape, which is maintained for about 0.15 s after the beginning of the gate motion. In order to analyze the shape of the wave head, the trajectories of the wave front and of three points characterized by water depths of 0.015 m, 0.02 m, and 0.03 m were measured. The paths of the wave front and of these points were simulated with the mathematical model (Fig. 9).

Fig. 8 - Wave profile

The computed values agree with the experimental results, excluding the path of the wave front close to the gate, where the flow is strongly perturbed and the measurments are more uncertain.

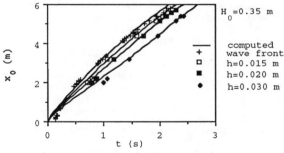

Fig. 9 - Path of the wave front

As the wave travels downstream, it seems to become more steep. In fact the time interval between the passing of the wave front and of the water depth h=0.02 m varies from about 0.25 s, when the wave front is 2 m downstream from the gate section, to about 0.18 s when the front is close to the downstream boundary.

Propagation on adverse slope
The experiments were conducted by using two values of the initial water depth upstream from the gate: H_0=0.20 m and H_0=0.30 m. The values of the adverse slopes were varied from S_0=-0.084 to S_0=-0.15.
The stage hydrograph at the section C, where the slope changed, and the trajectory of the wave front were measured. The most significant results are given in

Fig. 10, which shows the positions of wave front x_C, measured from section C, versus time.

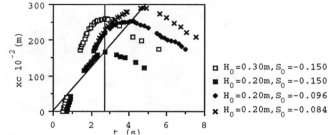

Fig. 10 - Distance travelled by the wave front on adverse slope

For the same slope, the distance travelled by the wave front depends on the initial water depth H_0. However the maximum distance for different values of H_0 is always reached at the same time.
On the other hand, considering the same value $H_0=0.20$ m but changing the slope, the distance routed by the front increases with the slope, as may be expected, but the maximum values are perfectly aligned. We conclude from this that the wave front travels on adverse slopes with an almost constant average celerity. This results are useful since they can be applied to different slope conditions. However, they need to be verified by considering a wide range of values for the adverse slope.
The maximum level H, above the horizontal bottom, reached by the wave front was measured and it is shown in Table II.

Table II: Maximum level reached by the wave front

H_0 (m)	S_0	H (m)
0.30	-0.150	0.394
0.20	-0.150	0.252
0.20	-0.096	0.247
0.20	-0.084	0.242

The specific head increases, as verified by De Marchi [2], who assumed that at the beginning of motion the total (potential) energy was completely transformed into kinetic energy. Assuming the water depth at the wave front to be zero, De Marchi found out that this total kinetic head is twice the initial energy. The kinetic energy changes back into potential energy, with losses, when wave front stops.

During the recession phase, the wave front became steeper (Figg. 11 and 12). This is probably beacuse it is supported by the rest of the wave which is still advancing.

Figg. 11, 12 - Wave profile on adverse slope

Reflection of the wave at the downstream boundary
Fig. 13 shows the reflected wave. Three different zones can be distinguished in the flow, which is very perturbed owing to vortexes which have the effects of trapping a large quantity of air into the water.

Fig. 13 - Wave reflection

The first zone, close to the bottom, consists only of water. The second zone is higly aerated. The third zone consists of foam where air prevails on water.
The water depth is initially higher than H_0, mainly due to a large quantity of foam. This foam tends to disappear and the stage stabilizes at a value slightly lower than H_0.
The injection of a tracer allowed to study in detail the mixing of flows caused by the superimposition of waves travelling in opposite directions due to reflection. In a section downstream from the gate, the trace is still oriented downstream even when the reflected wave is passing above. This shows that the two flows are completely separated (Fig. 14).

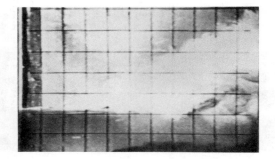

Fig. 14 - Flows travelling in opposite directions

Gradually the flows begin to mix in the middle of the section. After some time, the trace shows a strong turbulence and is nearly vertical. This indicates a condition of stagnation, and the flows are now completely mixed.
When the reflected wave at the upstream flume boundary induces a decrease in water depth and momentum, then the surge moves upstream.
Therefore the passing of the reflected wave is not followed by the immediate mix of the flows: the assumption that the water velocity is uniformly distributed immediately upstream and downstream from the surge does not seem to be experimentally verified.

CONCLUSIONS

Investigative experiments were conducted by removing a top hinged sluice gate which dams a rectangular flume. The experiments were carried out with the aim of studying in detail the flow through the gate section and the celerity of propagation of the wave after the rapid lifting of the gate.
In the gate section, the discharge presents a clear peak immediately after opening. This maximum value is greater than the analytical value of about 30%. When maximum discharge occurs, the celerity of the wave front propagation shows a pronounced peak, decreasing as the wave propagates downstream.
Wave propagation experiments on adverse slope showed that the level reached by the flow, measured above the horizontal bottom, is higher than the initial depth upstream from the gate, thus verifying an increase in the specific energy.
Other tests were conducted by imposing a wave reflecting wall at the downstream section of the flume and injecting a tracer. Even when the reflected wave

is passing upstream, the two flows are completely separated and travel in opposite directions. Only after a significant time period the flows are completely mixed.

REFERENCES

1   Chen, C.L., Laboratory Verification of a Dam-Break Flood Model, ASCE, Journal of Hydr. Div., 106, HY4: 535-556, 1980.
2   De Marchi, G. Onde di depressione provocate da apertura di paratoia in un canale indefinito, L'Energia Elettrica, XXII, 1945.
3   Dressler, R.F. Comparison of Theories and Experiments for the Hydraulic Dam-Break Wave, Int. Soc. of Scient. Hydrol., 101, 38: 319-328, 1954.
4   Faure, J., Nahas, N., Etude numerique et experimentale d'intumescences a forte courbure du front. La Houille Blanche, 5: 576-587, 1961.
5   Montuori, C. Immissione di una portata costante in un canale vuoto, Proc. Fondazione Politecnica per il Mezzogiorno, VI: 1-59, 1964
6   Montuori C., Introduction d'un Debit Constant dans un Canal Vide, Proc. XI Congress of I.A.H.R., Leningrad, 1965
7   Natale, N., Savi, F. Propagazione di onde di sommersione in un canale vuoto, Proc. of Jornadas de encuentro trilateral para el estudio de la hidraulica de las ondas de submersion, Zaragoza, 46-77, 1992.
8   Schoklitsch, A. Experiments on Flood Wave caused by Breaking of Dam, Hydraulic Laboratory Practice, 315-325, 1926.
9   Su, S.T., Barnes, A.H. Geometric and Frictional Effects on Sudden Realese, ASCE, Journal of Hydr. Div., 96, HY11: 2185-2199, 1970.
10 Trifonov, E. K. Experimental investigation of Positive Waves Propagation along Dry Bottom, Trans. Scient. Research Inst. of Hydrotecnics, Leningrad, X, 1933.
11 Whitham, G.B. The effects of Hydraulic Resistence in the Dam-Break Problem, Royal Society of London, Proceeding Series A, Mathematical an Physical Science, 226: 399-407, 1955.
12 Yevjevich, V. Sudden Water Release, Unsteady Flow in Open Channels, Mahamood and Yevjevich Ed., Fort Collins, 587-668, 1975.

Free surface flow modelling on a complex topography

Mohamed Naaim and Gerard Brugnot
Division Nivologie - CEMAGREF
BP : 76 Domiane universitaire
38402 Saint Martin d'Hères
France
E_mail : naaim@cmgr01.grenet.fr

Abstract :

The aim of this paper is to describe the different construction stages and the validation of a free surface flow model on a complex topography. This model is based on the two dimensional shallow water equations. In order to treat the complex topography a finite element space description is used. It allows us to follow the domain boundary and to use a reduced cell number. To solve these equations two numerical methods are used. The first one is based on a finite elements formulation. The Galerkin formulation is modified to add two new terms called stream line diffusion and shock capturing. These two terms make the method steady for the hyperbolic non-linear system. The second one is based on a finite volumes formulation. It consists in integrating the system on each cell and in applying a Riemann solver to obtain the numerical flux at the cell interface. The speed and the accuracy of the two methods are compared in the case of a dam break on an initially dry bottom. The accuracy obtained through the two methods is the same. Concerning the speed, the finite volumes method is more effective, therefore it was adopted. The model developed in this way was tested comparing it to experimental results obtained in the cases of a two dimensional dam break and a wave propagation in a lake with a complex topography. The obtained results show a good agreement between the numerical simulation and the experimental results.

1. Introduction

In the mountain areas we observe two naturals hazards, whose damages are produced by free surface flows. The first one is the case of the waves generated in a lake by the impact of a landslide. The second one is the flow resulting from a dam break. As flows occur on a complex topography it adds another difficulty to the resolution problem.

If it's exact that the creation of the wave in a lake by the impact of a landslide is a complex and three dimensional phenomenon, the experience shows that the resulting waves are often one dimensional in the direction of the landslide impact velocity. Therefore it's possible to study the generation of the wave through a one dimensional approach. The interactions between the wave and the topography produce a complex flow. It's impossible to study the propagation of the wave with a one dimensional model.

In the case of a dam break, two kinds of flows are possible. If the flow is sufficiently channelised, it's possible to use a one dimensional model. If the flow is more complex it's necessary to develop a new model integrating a two dimensional flow description and the interaction with the topography.

The present work objective is to develop a free surface numerical model based on the two dimensional Saint Venant equations on a complex topography. It proposes two numerical approaches using an unstructured space finite elements grid. The first is a generalisation of finite volumes Godounov schemes and the second uses a new finite elements formulation. The Galerkin formulation is modified by adding to it a new term in order to make the formulation more stable in the case of an non linear hyperbolic system.

2. Adaptative mesh refinement

The sensitivity, to the space discretisation, of the Godounov numerical scheme applied to Saint Venant equations was analysed [Naaim91]. It shows that the space steps must be adapted to the slope angle. If it is not the case parasite waves are born near the slope angle changes. These waves propagate and disturb all the flow. It indicates that the space steps must be small where the slope angle is high. The application of this criterion in the case of finite difference grid implies a small space step and a large cell number.

Three imperatives force to use a finite elements grid. It allows to adapt the space step to the slope angle without increasing the cell number, to use a grid composed by triangles and quadrangles and to follow the complex flow domain boundaries.

Along the contour lines the slope angle is equal to zero. These lines are drawn nearer in high slope angle zones and further in the low slope angle zones. It is logical to build the grid on the contour lines and to impose the space step corresponding to the local slope angle. This way, we obtain a decomposition of the flow zone into domains with more or less complex shapes. Each complex domain is decomposed into smaller domains with a more simple geometry. We use a program to obtain a finite element grid out of each small domain. Then the grids are merged together to obtain the whole grid.

3. Model formulation

The free surface flow dynamics is governed by the Saint Venant equations. They are obtained from Navier Stokes equations using the following hypothesis :

- incompressible fluid,
- hydro-static pressure distribution,
- the friction and the turbulence diffusion are represented by Chezy formula.

The conservative form of these equations is given by :

$$\frac{\partial U}{\partial t} + \text{div}(F) = S(U, \bar{x}, t) \tag{1}$$

$$U = (h, h\bar{u}), \quad F(U) = \begin{bmatrix} hu & hu^2 + \frac{1}{2}gh^2 & huv \\ hv & huv & hv^2 + \frac{1}{2}gh^2 \end{bmatrix}^t$$

and $\quad S(U) = \left[0, g\bar{\nabla}(hz_f) - \frac{\|\bar{u}\|}{C^2 h}\bar{u} \right]^t$

$\bar{x}$ is the space coordinate, t is the time coordinate , $\bar{u}$ is the velocity, h is the water depth, z_f is the level of the bottom, C is the Chezy coefficient and g is the acceleration due to the gravity.

In the one dimensional case, the system can be written in the following form :

$$\frac{\partial}{\partial t}\begin{bmatrix} h \\ hu \end{bmatrix} + \frac{\partial}{\partial x}\begin{bmatrix} hu \\ hu^2 + \frac{1}{2}gh^2 \end{bmatrix} = \frac{\partial U}{\partial t} + \frac{\partial f}{\partial x} = 0$$

The speed of the gravity wave and the Jacobean of f are respectively given by :

$$c = \sqrt{gh}, \quad \frac{\partial f}{\partial U} = \begin{bmatrix} 0 & 1 \\ c^2 - u^2 & 2u \end{bmatrix}$$

For h>0, the system is strictly hyperbolic; it possesses two eingenvalues (λ_1=u-c and λ_2=u+c) associated to two eingenvectors ($\bar{r}_1$ and $\bar{r}_2$). The two eingenvectors are given by :

$$\bar{r}_1 = \frac{1}{\sqrt{1+(u-c)^2}} \begin{bmatrix} 1 \\ u-c \end{bmatrix} \text{ and } \bar{r}_2 = \frac{1}{\sqrt{1+(u+c)^2}} \begin{bmatrix} 1 \\ u+c \end{bmatrix}.$$

The Saint Venant system has a pair of Riemann invariants witch satisfy :

$$\bar{\nabla} w_1 . \bar{r}_1 = 0$$
and $\quad \bar{\nabla} w_2 . \bar{r}_2 = 0.$

Two possible expressions of w_1 and w_2 can be given by :

$w_1 = u+2c$
and $\quad w_2 = u-2c.$

Thanks to the Riemann invariants property the system without second member can be written in a diagonal form :

$$\frac{\partial}{\partial t} \begin{bmatrix} w_1 \\ w_2 \end{bmatrix} + \begin{bmatrix} \lambda_1 & 0 \\ 0 & \lambda_2 \end{bmatrix} \frac{\partial}{\partial x} \begin{bmatrix} w_1 \\ w_2 \end{bmatrix} = 0$$

In the smooth zones this form permits to solve the system. It is the characteristic method. The system is non linear. Two different characteristics issued from two different points can collide making impossible the existence of smooth solutions (appearance of a discontinuity). Therefore, it is necessary to search the solution among the weak solutions of the system checking by :

$$\forall \varphi \in RxR^+ \int_{RxR^+} (U\frac{\partial \varphi}{\partial t} + F(U)\frac{\partial \varphi}{\partial x})dxdt + \int_R \varphi(x,0)U_0(x)dx = 0$$

It's possible to demonstrate that this formulation is compatible with hydraulic jump equations.

The system possesses a strict convex entropy : $\eta = \frac{1}{2}(hu^2 + gh^2)$ and two important properties also : symetry and invariance by rotation [Naaim, Vila, 90].

4 Numerical methods

The free surface flows have been for a long time treated by the characteristic method. This method is not capable to simulate flows with a discontinuity. In spite of this, many authors continue to use it. They separate the smooth zone and the discontinuous zone for the treatment.

The recent progress realised in the numerical analysis of the non linear hyperbolic systems allows to remedy to this disadvantage. The new schemes are able to treat the discontinuous points as the remaining of the flow. Two new methods are presented and compared in the following section.

4.1 Finite elements method

The classical Galerkin formulation corresponds to central scheme. This one doesn't grant any importance to the flow direction. In the case of non linear hyperbolic system this scheme is not stable. It is possible to remedy to this problem. If we use the upwind technic we reinforce the diagonal of the system and it becomes stable. Different upwind methods exist. The one proposed by Seppezy is used here. It consists in adding new terms to the Galerkin formulation.

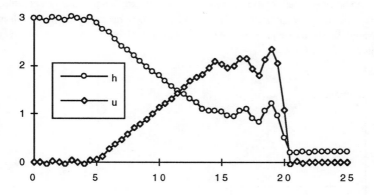

Figure 1: dam break simulated by Galerkin formulation

The mathematical environment is introduced. The time is discretised by a series t_n. Denote $I_n=[t_n,t_{n+1}]$ and $\Omega_n=R^m x I_n$. For h>0 and n>0, T^h is a triangulation of Ω composed of a quadrangles and triangles who have h as diameter.

Denote $\psi_h^n=\{\psi \in [H_1(\Omega_n)]; \psi_{/K} \in P_k(K), K \in T_h^n)\}$ where $P_k(K)$ is the set of polynoms so that the degree is less than k and $H_1(\Omega_n)$ is the set of the functions carresommables over Ω_n.

The stream line diffusion and shock capturing formulation proposed by Seppezy define the approximate solution as a function in ψ_h^n.

Let $\quad L(U) = \dfrac{\partial U}{\partial t} + \dfrac{\partial F(U)}{\partial x}.$

For $\forall\ \varphi \in [\Psi_h^n]$:

$$\int_\Omega L(U).(\varphi + \tau(\frac{\partial \varphi}{\partial t} + \frac{\partial F(U)}{\partial U}\frac{\partial \varphi}{\partial x}))dv + \int_\Omega \varepsilon_1(U).\nabla U.\nabla \varphi.dv + \int_\Omega \varepsilon_2(U).\frac{\partial U}{\partial x}.\frac{\partial \varphi}{\partial x}.dv = 0$$

Where
$$\varepsilon_1(U) = \delta \frac{\int_K |L(U)|(1 + \left|\frac{\partial F(U)}{\partial U}\right|)dv}{\int_K dv}(1 + \sum_{|\alpha|=3}^{k+2} \|U\|_{L\infty}^{\alpha-1} \cdot \|D^{\alpha}\eta(U)\|_{L\infty(K)})$$

and
$$\varepsilon_2(U) = \gamma \frac{\int_{K \cap I_n} |U^+ - U^-|dx}{\int_{K \cap I_n} dx}(1 + \sum_{|\alpha|=3}^{k+2} \|U\|_{L\infty}^{\alpha-1} \cdot \|D^{\alpha}\eta(U)\|_{L\infty(K)})$$

Where $D^{\alpha}\eta(U)$ represents the h(U) derivative using the multi-indexed notation. τ, δ and γ are parameters to be determined for each system. The upwind term is represented by τ multiplied by the hyperbolic operator applied to φ. The shock capturing terms are represented by δ and γ. They are multiplied by a term proportional to the solution residue for δ and to the solution jump at the boundaries of I_n for γ.

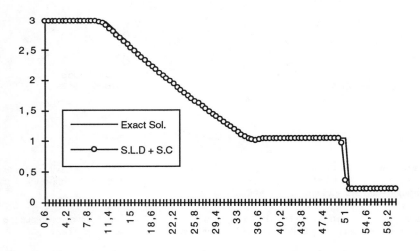

Figure 2 : Dam break simulated by the SLD finite element method

4.2 Finite volumes method

In the case of finite volumes method the goal is to approach U by its mean value and its gradient in each cell K_i. The mean value is obtained at t^n (res. à t^{n+1}) by :

$$U^n = \int_{k_i} U(\bar{x}, t^n)ds$$

and
$$U^{n+1} = \int_{k_i} U(\bar{x}, t^{n+1})ds$$

The gradient on each cell is obtained in two steps. In the prediction step the mean gradient is obtained by using the neighbour U cell value. To make the numerical scheme stable it is necessary to limit the obtained gradients. The limitation criterion is that the gradient module must be reduced if it creates a new local maximum or new local minimum. This criterion is applied for all the cell boundaries. For each one denoted Γ_{ij} the U value obtained at the centre of the edge must be in the interval determined by U_i and U_j.

The integration of the system on each elementary volume $V_i = K_i \times I_n$ gives :

$$\iint\limits_{v_i} (\frac{\partial U}{\partial t} + \mathrm{div}(F))\,ds\,dt = \iint\limits_{v_i} S\,ds\,dt$$

The application of the Green formula followed by the application of the mean formula gives :

$$\exists t_s \in [t^n, t^{n+1}], \exists x_s \in \Gamma_{ij} \,/\, U^{n+1} = U^n - \frac{\Delta t}{S} \sum_j \int_{\Gamma_{ij}} F(U(\bar{x}_{sj}, t_s)).\bar{n}.\,d\Gamma + \Delta t G(U_i(t_s))$$

Where $\bar{n}$ is the external normal to the K_i boundary Γ_i. A numerical approximation chosen is to take $t_s = t^n$ (explicit scheme) and the numerical flux is calculated in the centre of the edge.

The scheme formulation is then :

$$U^{n+1} = U^n - \frac{\Delta t}{S} \sum_j \int_{\Gamma_{ij}} F(U^n(\bar{x}_c)).\bar{n}.\,d\Gamma + \Delta t G(U_i^n)$$

It remains the calculation of the numerical flux on each edge boundary of K_i . In the case of the one dimensional problem, an exact calculation is possible using a Riemann's problem solver [Vila 86]. In order to reduce the flux calculation to a one dimensional case we proceed in two steps. The Saint Venant system is written in the local coordinate system $(\bar{n}, \bar{\tau})$ formed by the tangent and the normal to the edge. Then we neglect the variation along the tangent direction $\bar{\tau}$. This way the problem becomes one dimensional and we can use the one dimensional Riemann's problem solver [Naaim 91].

The Riemann's problem is defined by :

$$\begin{cases} \dfrac{\partial U}{\partial t} + \dfrac{\partial f}{\partial x} = 0 \\ U(x,0) = \begin{cases} U_g \text{ if } x < 0 \\ U_d \text{ if } x > 0 \end{cases} \end{cases}$$

and for each Γ_{ij} edge the U left and right value are given by :

$$\begin{cases} U_g = U_i + \bar{\nabla}U_i.(\bar{x}_\Gamma - \bar{x}_i) \\ U_d = U_j + \bar{\nabla}U_j.(\bar{x}_\Gamma - \bar{x}_j) \end{cases}$$

This scheme is stable if the Courant number formed is less than 1.

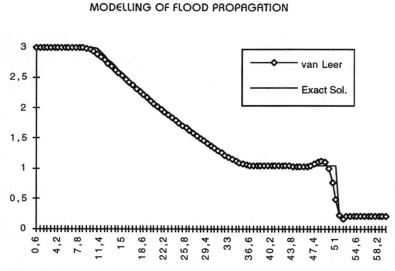

Figure 3 : Dam break simulated by the van Leer scheme (compared to the exact solution)
(result obtained in the case of diagonal propagation of a dam break on regular grid)

5. Comparisons finite elements / finite volumes

In this section we study the accuracy and the speed of the finite elements and finite volumes
schemes. We remember that for the smooth solution the accuracy rate of the finite elements
method is linked to the degree of the polynoms used in the formulation and that the van Leer
scheme's one is 2. This is not true in the case of discontinuous solution. An empirical method
is used below to determine the accuracy rate for the studied schemes. The exact solution of a
dam break on horizontal bottom is used. The two schemes are used to reproduce this solution
for many space steps.

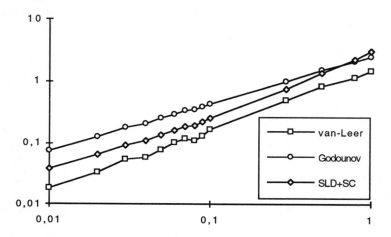

Figure 4 : Error function of space step h for different numerical schemes

The error between the exact and simulate solution is calculated by : $E_r = \int_0^L |u_s - u_{ex}| dx$. It's approximated by $E_r = A.h^\alpha$. The obtained value for α are put in the following table :

Scheme	α	CPU (Time)
Godounov	0,82	1
SLD (first order)	0,9	420
van Leer (Second order)	0,95	1,8

The van Leer scheme and the stream line diffusion formulation (k=1) give the same accuracy. The disadvantage of the second method is its expensive cost. It requires a large computer memory capacity. It takes two hundred times more time than with the van Leer scheme. These two disadvantages suggest to choose the van Leer scheme and to conclude that the stream line diffusion is not adapted to study the rapid transient flows.

6. Model validation and conclusions

The validation of the model will be done on two cases. The first treats of a two dimensional dam break on sudden enlargement and the second treats of the propagation of a wave on a complex topography.

6.1 Dam break on a sudden enlargement

The bibliography existing on the small scale model concerning dam break is abundant (see references). However, an analysis from this bibliographical research shows a few analyses on flow front propagation into sudden enlargement. Because of this deficiency, one experimental study has been carried out at CEMAGREF using image processing techniques : different steps of the front propagation in a channel have been studied.

The CEMAGREF experimental channel is : 8m (long) x 1m (wide) x 0.6 m (depth). A reservoir (1 m wide and 1 m longitude) has been constructed in the upstream zone. From its right side a fast aperture water-gate has been set up to simulate the dam break. The measurements have been made using capacity sensors at the internal and the external sides of the dam and an imaging device in the down stream part. The water height before the break (spelled h_0) is measured using an adhesive tape into the reservoir. For each height h_0, the experiments are realised in two stages. At first, using plan water for height measurements during the break, in the internal and external dam sides. A second, allows to follow the front evolution through the use of coloured water, for a better contrast. This experimental set is used to validate the presented model.

A two dimensional mesh for the channel is realized. Different experiements were realized using different initial water depths. Then, the measured water height inside of the reservoir is compared with the numerical simulated one. The front dynamics outside the reservoir is studied experimentally and numerically. As the front has a two dimensional extension, we compare the front position coordinates along 4 directions. These directions are distributed regularly on the sudden enlargement. The first one is orthogonal to the gate. The figure 6 shows the comparison between the water height in the reservoir measured and simulated in the case of $h_0=0.6$ m. The figures 7 and 8 show the comparison between the measured covered distance and the simulated one corresponding to two differents directions.

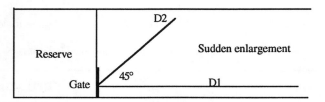

Figure 5 : Sketch of the device

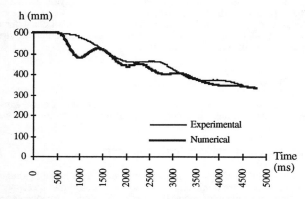

Figure 6 : Height measured and calculated inside the reserve

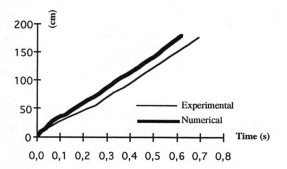

Figure 7 : Distance covred (cm) in time (s) by the front on D2 Direction

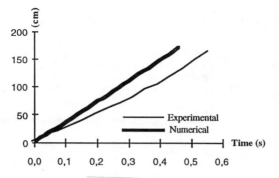

Figure 8 : Distance covred (cm) in time (s) by the front on D1 Direction

An important aspect of the flow at the sudden enlargement is the formation of a hydraulic jump. It is formed when the flow is reflected by the channel left wall : it has a triangular shape. It is observed but it is not possible to make measurements. The numerical model reproduces correctly the shape and the location of the jump.

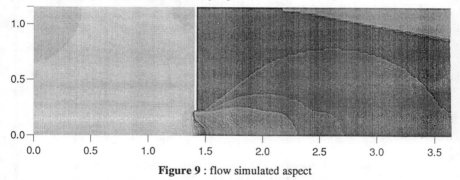

Figure 9 : flow simulated aspect

From the comparisons between experimental and numerical results, we can conclude that the model is capable to simulate the reservoir emptying, the flow front dynamic in the sudden enlargement and the location and the shape of the hydraulic jump.

6.2 Wave propagation on a complex topography

The Dillan landslide on the NW slope of Grand Maison reservoir, had been identified during the spring of 1986 by cracks apparitions. This landslide constitutes a risk for the dam because of possible waves production. The small scale model has been constructed by the LNH, and the study of landslide consequences in the reserve carried out. A lot of data were got from the instrumental small scale model. Theses measurements have been used to test the model of wave propagation on a complex topography.

In the reservoir, the water levels were registered by capacity sensors. The piston effect, originated by the landslide, produces a negative wave on the concerned slope and a positive wave on the opposite slope. Theses initial perturbations are followed by secondary wave groups. The wave propagation goes with reflections and refractions due to rough terrain.

The simulation of waves propagation by the model has been carried out with a special grid for a part of the reservoir. This grid is adapted to the topography. It's leaned upon four contourlines representing the relief.

Then, model initialisation is done from the steady solution consisting in a horizontal water surface with null velocities. Thus, the created wave due to the landslide is recorded by the sensors and is used for the upstream limit condition in the model. Afterwards, we have compared both waves : simulated propagation and empirical propagation, considering punctual and two-dimensional aspects.

The obtained results show a good agreement between the numerical simulation and the experimental measures.

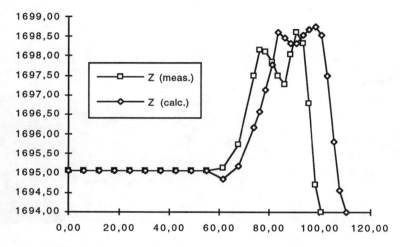

Figure 10 : Comparison between the simulated and measured wave in GrandMaison reserve (z (m) function of time (s))

References :

Bell S. W., Elliot R. C., Chaudry M. H., "Experiemental results of two dimensional dam break flows", Journal of Hydraulic Research, 1992, Vol 30, n° 2, p 225-252.
Bellos C. V., Soulis J. V., Sakkas J. G., "Experimental investigation of two dimensional dam break induced flows", Journal of Hydraulic Research, 1992, vol 30, n° 1, p 47-63
Godounov, S.K, "A difference method for numerical calculation of discontinuous eaqutions of hydrodynamics", Math Sb, 47 (89), 1959, 271-300.
Naaim M., "Modelisation numerique des effets hydrodynamique provoqués par un glissement solide dans une retenue", Thèse de l'université Joseph Fourier de Grenoble, 1991.
Naaim, M, Vila , J.P, "Analyses des effets provoqués par la chute d'une avalanche dans une retenue", rapport final de la convention CEMAGREF/EDF (1989), 64 pages.
Szepessy A., "A streamline diffusion finite element method for conservation laws", Goteborg university thesis.
Vila, JP, "Simplified Godounov schemes for 2x2 systems of conservation laws", SIAM Num. An., 1986, 23, 6 1173-1192.

Two-Dimensional Modelling of Flow in the River Sava

Rudi Rajar and Matjaž Četina
University of Ljubljana, FAGG
Hajdrihova 28
61000 Ljubljana, SLOVENIA

ABSTRACT

On the river Sava near Ljubljana, a sports centre with a kayak racing channel was built in 1948. For the world championship in kayak-canoe racing in 1990 totaly new racing channel had to be built, since the old one did not correspond to the international racing norms.

The planned reconstruction of the channel, with several new constructions (stands, platforms) demanded answers of the following questions:
1. How will the changed topography influence the water levels at high water flows (flooding of some houses along the banks was in question).
2. How will the discharge over a weir be distributed between the racing channel and the main river channel.
3. How to design the new racing channel to get the optimum performance.

All three problems were solved by a 2D depth-averaged mathematical model. Basic equations were used in the conservative form. They were solved by a finite-volume numerical method. A combined central-differences and upstream differences were used and an implicit iterative scheme was applied for the solution.

The model showed very good performance even for the simulation of the highly turbulent flow in the racing channel.

Only the first problem is described more in detail. References are given for the modelling of the flow in the racing channel.

INTRODUCTION

On the river Sava, about 10 km NW of Ljubljana, there is a sports centre, where a kayak racing channel was built in 1948. Fig. 1 shows the general situation of the region. There is an oblique weir upstream of the region, and below it an island (which is totally flooded at high water flows) separates the flow into the racing channel (right) and into the river Sava main channel (left). On the right bank, there is a small hydroelectric power plant (HEPP) which conducts about 25 m^3/s over its headrace and tailrace channels.

A world championship was planned in 1990, but the existing channel did not correspond to the international norms for kayak-canoe racing. Therefore a reconstruction of the racing channel was planned, which should be longer, narrower and mainly more difficult. The building of several new constructions (stands, observation towers, platforms) was also planned along the new channel. So the topography of the terrain along the banks was to be changed and also one part of the river cross-section and the bottom topography at the right bank.

Therefore one of the questions to be solved by mathematical modelling was, how the changed river topography would influence the water levels in the river at high water flows. Along the right bank there are some houses, which had already been slightly flooded during hundred-year discharges, and concern existed that the new topography might raise the water levels significantly.

Another question to be solved by the mathematical model was, how the discharge over the weir (Fig. 1) would be divided between the racing channel and the river Sava. This would give us the relation between the total discharge in the river and the discharge in the racing channel. The discharge in the river can be regulated exactly at an upstream HEPP (Medvode) and with the help of the above mentioned relation the necessary discharge in the racing channel can be maintained during the races.

Because of the unusual, broken form of the weir and because of the very complex topography of the river and the two islands in this region, a 2D model had to be used. A model, developed in 1988 (Četina [2]) and used already for several practical applications (Rajar and Četina [6], Četina [3]) was used. Calculations were also executed by a 1D model for comparison purposes.

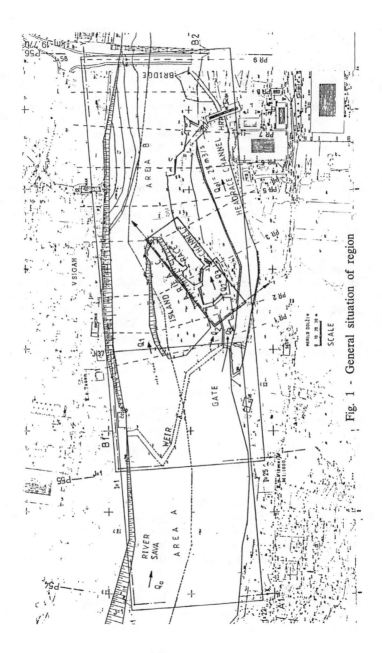

Fig. 1 - General situation of region

INPUT DATA

Computational Area
We divided the computational area into two regions, shown in Fig.
1:
1. Region A - upstream of the weir of length app. 300 metres.
2. Region B - area downstream of the weir, comprising the
   racing channel, the island, the Sava main channel and the
   river section downstream to the road bridge at Tacen. The
   whole region B is about 500 metres long.

The simulation was performed separately for each region, because
critical conditions exist on the weir at both low and high
discharges. At low discharges Region A was simulated to determine
the distribution of the discharges over the weir. At high waters the
main results to be determined were water levels in Region B.

Topographic Data
The geometry of the region was evaluated on the basis of the
cross sections whose situation is shown in Fig. 1 and after a
detailed measurement of the river bottom, after which a detailed
map with bottom levels was made. As for the 2D modelling we
need the bottom levels at all the points of the relatively dense
numerical grid, interpolations were made for these points from the
map.

A 1D simulation of the downstream Region B was also made and
the cross-sections 1 to 9 were used for evaluation of the
topography.

The topography of the final, changed configuration was evaluated
from the design plans.

Hydrological Data
These data were obtained from a study of Sava hydrology dating
from 1983. Since the measuring station is about 8 km downstream,
the discharges were reduced with a factor of 0.959 in relation to
both watersheds. The high water discharges are: $Q_{10}=1180$ m^3/s,
$Q_{50}=1593$ m^3/s and $Q_{100}=1708$ m^3/s.

Hydraulic Data
The stage-discharge relationship of the downstream section at the
Tacen bridge was determined on the basis of the registered water
elevation at the bridge piers at high water on Jan. 1979. For other

discharges the relationship was extrapolated on the basis of the Manning's formula.

The roughness coefficient should be known at every point of the numerical grid. In Region A Manning's roughness coefficient was estimated to be $n = 0.03$ sm$^{-1/3}$. In Region B the bottom roughness varies very much. We determined the coefficients from 1:1000 and 1:200 situations and from the survey of the terrain. Final values were obtained on the basis of the calibration. They varied from 0.025 to 0.15, the highest values were found for the flooded island, where brush and small trees cause very high flow resistance. For the planned configuration (after the reconstruction) the roughness coefficient was diminished to 0.030 in the region of the new, mainly concrete structures along the racing channel.

Discharge through the Hydroelectric power plant. The flow through the headrace channel of the HEPP was eliminated from the simulations, taking into account that the nominal discharge of 25 m^3/s leaves the computational region at the entrance of the channel and reenters Region B as side inflow (Fig. 1).

The discharge over the weir is needed as the downstream boundary condition for Region A. We took into account the equation for a broad-crested weir (Agroskin [1]):

$$Q = \varepsilon m l \sqrt{2g} H_0^{3/2} \tag{1}$$

where l = overall length of the weir, H_0 = energy elevation over the weir, g = acceleration due to gravity, m = weir coefficient (with a value of 0.32), $\varepsilon = 1.0$ since the height of water over the weir (app. 1 m) is much smaller than the weir length (app. 100 m). The overall length of the weir was taken into account, with the velocity direction being normal to the weir crest. By observations in nature it was found that this was generally true, the velocity directions being much closer to the perpendicular direction than to the direction of the flow just upstream of the weir.

In the 2D model we need the relation between the depth-averaged velocity (V) and the water depth (h) in every grid cell on the weir. Therefore Eq. 1 was transformed into:

$$V = \left[m\sqrt{2g}\,(h + \frac{u^2 + v^2}{2g})^{\frac{3}{2}} \right] / h \qquad (2)$$

The velocity V was decomposed into two components parallel to the two axes of the numerical grid.

DESCRIPTION OF THE TWO-DIMENSIONAL MATHEMATICAL MODEL

Basic Equations
Equations of 2D depth-averaged steady flow were used (Četina [2]):

$$\frac{\partial h}{\partial t} + \frac{\partial (hu)}{\partial x} + \frac{\partial (hv)}{\partial y} + q = 0 \qquad (1)$$

$$\frac{\partial (hu)}{\partial t} + \frac{(hu^2)}{\partial x} + \frac{\partial (huv)}{\partial y} = -gh\frac{\partial h}{\partial x} - gh\frac{\partial z_b}{\partial x} - ghn^2 \frac{u\sqrt{u^2 + v^2}}{h^{4/3}} + \frac{\partial}{\partial x}(hN\frac{\partial u}{\partial x}) + \frac{\partial}{\partial y}(hN\frac{\partial u}{\partial y}) \qquad (2)$$

$$\frac{\partial (hv)}{\partial t} + \frac{(huv)}{\partial x} + \frac{\partial (hv^2)}{\partial y} = -gh\frac{\partial h}{\partial y} - gh\frac{\partial z_b}{\partial y} - ghn^2 \frac{v\sqrt{u^2 + v^2}}{h^{4/3}} + \frac{\partial}{\partial x}(hN\frac{\partial v}{\partial x}) + \frac{\partial}{\partial y}(hN\frac{\partial v}{\partial y}) \qquad (3)$$

where: h =water depth, u and v =depth averaged components of velocity in the direction of the numerical grid axes x and y, n =Manning's roughness coefficient, z_b =bottom elevation, q =inflow/outflow from the cell (outflow as positive), N =diffusion coefficient.

Eq. 3 is a continuity equation, Eq. 4 and 5 are momentum equations for x and y directions. The last two terms in the momentum equations express the influence of the turbulent shear stresses between the cells. The model is completed with the known k-ε turbulence closure model, where the local values of the diffusion coefficient are computed at every grid point. But in this case numerical simulations showed that the influence of bottom roughness absolutely prevails over the turbulent shear stress between the cells. Therefore the last two terms of the momentum equations were neglected in the final computations. Since in the case described the flow was steady, the time derivatives in Eqs. 3 to 5 were zero.

Numerical Method

The equations were solved by a control-volume method of Patankar and Spalding (Patankar [5]). The method belongs to the group of finite-difference methods. The basic characteristics of the method are: a staggered grid, a hybrid scheme (combination of central-difference and upwind schemes) and an iterative procedure for the computation of pressure corrections. In the case of unsteady flow simulation a fully implicit scheme is used.

The numerical grid of Region A has 22×37 points. The space step in the x-direction is uniform ($\Delta x=7.5$ m), in the y-direction the step changes from 7.5 to 15.0 m. Region B is covered by a grid of 25×58 points with the same values of both space steps.

Boundary Conditions

The equations are elliptic, and therefore the boundary conditions should be determined along the whole (closed) boundary of the region.

Region A: At the upstream boundary the corresponding discharge was distributed uniformly over the inflow cross-section. At the downstream boundary (at the weir) the relation $V(h)$ already described for flow over the weir was taken into account. Additionally, outflow from the computational field into the racing channel (also the $V(h)$ relation obtained from the weir equation) and into the headrace channel of the HEPP ($Q=25$ m^3/s) was accounted for. At all the closed boundaries (banks) zero normal velocities were accounted for, and the gradient of longitudinal velocities normal to the boundaries was taken to be zero.

Region B: At the downstream boundary (Tacen road bridge) a $Q(h)$ relation was used. At the upstream boundary the distribution of the inflow discharge was obtained from the computations in Region A (flow over the weir). At the outflow from the tailrace channel of the HEPP an additional inflow into the computational field was accounted for.

Calibration of the Model and Accuracy of the Results

No systematically measured water levels were available. The model was calibrated on the basis of observed water levels on Jan. 28. and 29. 1979 with the discharge being 1228 m^3/s.

The results of the calibration are presented in Fig 4. The bottom roughness coefficients were varied to obtain the best agreement. As we see from Fig. 4, at one point (Breznik) the agreement is perfect. At the other two points the measured levels are about 0.5 m higher than the computed ones. But since at these two points the water level recordings were made outside the river channel, the velocity head $v^2/2g$ was mainly transformed into a water surface elevation. If we subtract these value (app. 0.3 m for the velocity of about 2.5 m/s) from the measured water levels, the difference between the simulated and the measured levels is only 0.2 m.

The agreement between the results of the 2D and the 1D model is relatively bad (the basic principles of the 1D model are described in the next section). The 1D model gives generally about 0.5 m higher elevations. The cause of the difference was found to be a hydraulic jump which appeared in the 1D computations because of too coarse a numerical grid (only 9 cross-sections). This hydraulic jump is not physically realistic and was not formed in the 2D computations where the numerical grid was much denser. But the overall pattern of the levels by both computations is parallel, which shows that elimination of the unrealistic hydraulic jump would give acceptable agreement.

The 2D model should be more accurate than the 1D one. But its accuracy is of course dependent on the accuracy of the data provided for the calibration and also for the final computations. Therefore we estimated that the possible discrepancies from the finally computed water levels can be approx. from -10 to +20 cm.

Although the computed absolute water levels are uncertain to the extent of the above mentioned values, we can positively assume that the computed changes of the water levels due to changes of the topography (which was the main goal of the study) are determined to an accuracy of about 5 cm.

DESCRIPTION OF THE 1D MODEL

Division of the flow in a given cross section was made by dividing the region into flow belts, each having different depth and different roughness. Transfer of the discharge between the belts was possible through the cross directional water surface differences. The method is especially applicable to prevalently 1D flow in rivers with important flood plains.

There is only one boundary condition needed in the simulation: the water level at the downstream boundary. Simulation of supercritical flow is not possible by the model described.

RESULTS AND CONCLUSIONS

<u>Region A - Determination of Discharge over the Weir.</u> Several simulations of low discharges (from 50 to 250 m³/s) in Region A and over the weir were performed. Figs. 2 and 3 show the computed velocity field for river discharges of $Q=50$ and 250 m³/s respectively. From field observations we know that at the point D (Fig. 2) the flow is divided: the part of the discharge which flows over the weir on the right hand side of this point, flows into the racing channel. The rest flows into the river Sava main channel (at low discharges left of the island). It was determined that the best racing conditions in the kayak racing channel are at a discharge of $Q=25$ m³/s, and the simulations showed that this is the case when the total discharge in the river Sava is 104 m³/s. The upstream HEPP Medvode can assure these conditions during the kayak-canoe races.

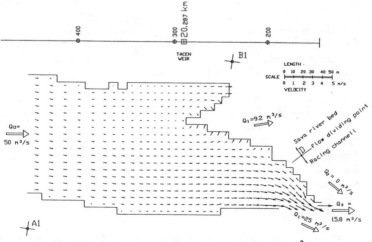

Fig.2-Region A - velocity vectors at $Q=50$ m³/s

<u>Region B - Water Levels at Flood Discharges</u>
Computed water levels and velocity fields for the topography before the changes are presented in Fig. 4 for the observed discharge of

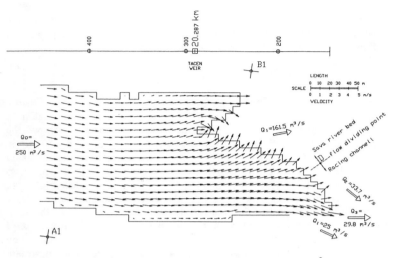

Fig.3-Region A - velocity vectors at $Q=250$ m^3 / s

1228 m^3/s and in Fig. 5 for the 100-year water flow of $Q=1708$ m^3/s. In Fig. 5 the water levels resulting from the new topography are also presented where they differ from the original ones. It can be seen that the differences are minimal: the planned changes in topography would cause a maximum increase of water levels of about 5 centimetres, which is practically negligible.

The velocity vectors at Q_{100} (Fig. 5) show that the main part of the discharge flows between the left bank and the island. Here are also found the maximum velocities (about 4.5 m/s). The island is than totally flooded and the water depth is up to 1.5 m. Past observations confirm this result. The velocities over the island do not exceed 1.5 m/s, due to the small depths and to very large roughness.

The main results of the 2D model simulations are:

1. During kayak races a total discharge of $Q=104$ m^3/s in the river Sava should be assured by the upstream HEPP, because this implies the discharge of $Q=25$ m^3/s in the racing channel. This discharge is needed for the best racing conditions.

MODELLING OF FLOOD PROPAGATION

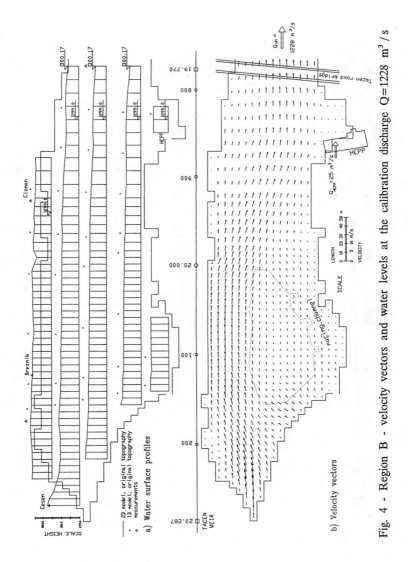

Fig. 4 - Region B - velocity vectors and water levels at the calibration discharge Q=1228 m³/s

a) Water surface profiles

b) Velocity vectors

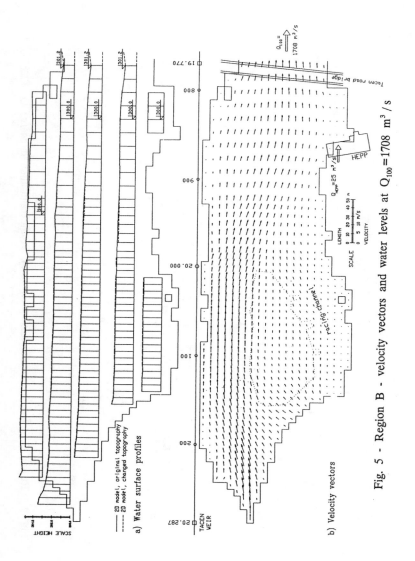

a) Water surface profiles

——— 2D model; original topography
- - - - 2D model; changed topography

b) Velocity vectors

Fig. 5 - Region B - velocity vectors and water levels at $Q_{100}=1708 \ m^3/s$

2. Simulation by a 2D mathematical model has shown that construction of the new racing channel with several accompanying structures (stands, observation towers, platforms etc.) will have only a negligible effect on the water levels at flood discharges (max. increase in the water levels of 5 cm for 100-year water levels). The conditions for several families living in the houses along the left bank of the river will not worsen.

SIMULATION OF FLOW IN THE KAYAK RACING CHANNEL

Another study was made to help in designing the new racing channel. We warned the designers that the mathematical simulation cannot give quite reliable results for such a complex flow with several hydraulic jumps, and transitions between the flow regimes. But due to lack of time and money it was not possible to build a physical model.

The same 2D model as described above was used and it was first calibrated and verified with measurements in the old racing channel. Since the flow was very turbulent, the terms expressing the turbulent shear stresses, together with the turbulence model k-ε, were taken into account. Unexpectedly the model showed the possibility to simulate most of the flow details relatively well. Since the momentum equations are in the sc. "conservative form", the energy losses, due to hydraulic jumps are also simulated relatively well. Transitions between supercritical and subcritical flow and vice-versa can also be simulated.

Therefore the flow in the first version of the new, projected racing channel was simulated. On the basis of the computed velocity field the designer changed some details of the banks and islands. After some trials the channel was built, but the form was not yet fixed in concrete. Experienced kayak canoeists competitors did some training in it and then proposed some additional changes. After that the final form was fixed.

During the world championship in 1990 the general opinion of the competitors was that the channel is very good and interesting, though a bit too difficult. Some minor changes were made later.

Fig. 6 shows the velocity field and Froud number distribution in the final channel. A detailed presentation of the whole research study is presented in Četina and Rajar [4].

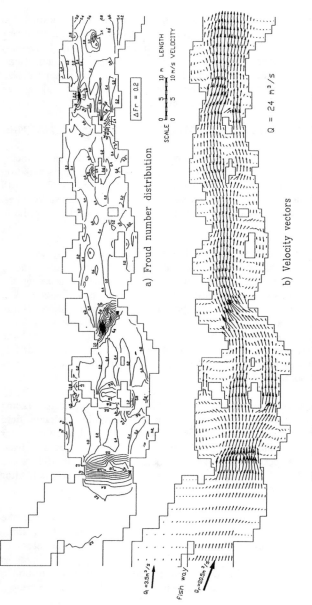

Fig. 6 - Froud number distribution and velocity field in the final channel

REFERENCES

1. Agroskin, I. I.: *Hydraulics,* Technical Publications, Zagreb, 1969.
2. Četina, M.: *Mathematical Modelling of 2D Turbulent Flows*, Masters degree Thesis, University of Ljubljana, 1988 (in Slovene).
3. Četina, M.: *Some Examples of 2D Simulations of Turbulent Free-Surface Flows,* Proc. of the Congress of the Yugoslav Association for Hydraulic Research, Sarajevo, 9.-13. 10. 1990, Proceedings, pp. 101-113 (in Slovene).
4. Četina, M. and Rajar, R.: *Mathematical Simulation of Flow in a Kayak Racing Channel,* Seminar on Refined Flow Modelling and Turbulence Measurements, Paris, 7.-10. 9. 1993, Proceedings, pp. 637-644.
5. Patankar, S.V.: *Numerical Heat Transfer and Fluid Flow,* McGraw Hill B.C, 1980.
6. Rajar, R. and Četina, M.: *Mathematical Simulation of 2D Lake Circulation,* HYDROSOFT Intern. Conference, Southampton, 9.-12. 9. 1986, Proceedings, pp. 125-133.

Dynamics of Sediment Transport

Flood Propagation on Mobile Beds under Mountainous Flow Conditions

C. Beffa and R. Faeh
Laboratory of Hydraulics, Hydrology and Glaciology (VAW)
ETH Zentrum, CH–8092 Zürich

ABSTRACT

An implicit finite-volume scheme is used to solve the shallow water equations under mountainous flow conditions. The algorithm is stable for sub- and supercritical flows and moving internal boundaries. Recent flood events in the Swiss Alps have shown the importance of morphological processes. A mobile bed module has, therefore, been implemented that accounts for suspended load and bedload. Numerical solutions are presented and compared with exact solutions and measured values. The paper is concluded with a practical application of the the model concerning a flood event in an alpine valley and a brief discussion of the results.

1 INTRODUCTION

Models for open channel flow and sediment transport present an advanced tool in hydraulic engineering. The increasing interest therein and the variety of applications has heightened the need for numerical models suitable to predict flows and sediment transport under mountainous conditions where steep slopes and small flow depths occur. Additionally, the internal boundaries (i.e. the boundaries between flooded and dry areas) are not known a priori or depend on the flow conditions. Thus, a numerical model must capture both transcritical flows (i.e. the transition from sub- to supercritical and vice versa) and moving internal boundaries in order to be a practicable tool for mountainous conditions.

Among the many numerical methods developed in attempting to solve the shallow water equations, the cell-corner scheme – or Preissmann scheme – and the staggered-grid scheme have become the standard methods for 1-d and 2-d calculations, respectively.[1,2] Because of their compactness and spatial accuracy these methods have been widely used in coastal and river applications. However, these methods do not converge under transcritical flow conditions. Stability is only achieved if the dynamic terms in the momentum equations are dropped.[3] Hence upwind schemes, which have been successfully applied for transonic flows in the field of aerodynamics, are proposed.[4, 5] Since these methods have to be expanded to higher accuracy they are not as compact as the former methods. This leads to relatively expensive models concerning implementation and CPU time. – Alternatively, a central scheme is proposed that is based on the Beam and Warming scheme for the Navier–Stokes equation.[6] It is stable under transcritical flow conditions while retaining the compactness of a central scheme. Of particular interest is the addition of artificial dissipation and the treatment of internal boundaries.

Presented problems, however, are seldom purely hydraulic in character. Recent flood events in the Swiss Alps have shown that extensive damage is caused by the entrainment and the deposition of sediments.[7] They are shifted from narrow and steep parts of a valley with high bed shear forces to the flat and wide valley bottom where the settlements are located. Here the principal damage occurs due to the flooding and the silting-up of the area. Flood propagation under these conditions has, therefore, to consider the morphological processes.

2 FLOW MODEL

The two-dimensional, depth-averaged shallow water equations can be written in Cartesian coordinates (x, y) as

$$u_{,t} + f_{,x} + g_{,y} + s = 0 \tag{1}$$

where

$$u = \begin{bmatrix} h \\ p \\ q \end{bmatrix}, \quad f = \begin{bmatrix} p \\ pu - \dfrac{h}{\varrho}\tau_{xx} \\ pv - \dfrac{h}{\varrho}\tau_{xy} \end{bmatrix}, \quad g = \begin{bmatrix} q \\ qu - \dfrac{h}{\varrho}\tau_{xy} \\ qv - \dfrac{h}{\varrho}\tau_{yy} \end{bmatrix}, \quad s = \begin{bmatrix} 0 \\ ghz_{,x} + \dfrac{\tau_{b_x}}{\varrho} \\ ghz_{,y} + \dfrac{\tau_{b_y}}{\varrho} \end{bmatrix}$$

with h=depth of flow, (p,q)=components of discharge per unit width, (u,v)=components of flow velocity, z=water level, τ_b=bed shear stress, g=gravitational acceleration and ϱ=density of water. The momentum equations are used in a non-conservation form for practical reasons. The bed shear stress is usually expressed by the quadratic friction law

$$\frac{\tau_{b_x}}{\varrho} = \frac{\bar{u}u}{c^2}, \quad \frac{\tau_{b_y}}{\varrho} = \frac{\bar{u}v}{c^2}$$

with the mean velocity $\bar{u} = \sqrt{u^2 + v^2}$ and the coefficient of the friction determined by the logarithmic formula or the Strickler formula

$$c = \frac{1}{\varkappa}\ln\left(11\frac{h}{k_s}\right) \qquad \text{resp.} \quad c = 7.6\left(\frac{h}{k_s}\right)^{1/6}$$

with $\varkappa$=Kármán constant and k_s=equivalent sand roughness. The turbulent shear stresses can be expressed by the Boussinesq relation

$$\frac{\tau_{xx}}{\varrho} = 2\nu_t u_{,x} \, , \qquad \frac{\tau_{yy}}{\varrho} = 2\nu_t v_{,y} \, , \qquad \frac{\tau_{xy}}{\varrho} = \nu_t \left(u_{,y} + v_{,x} \right)$$

The contribution of the bed friction to the turbulent viscosity is

$$\nu_t = \frac{\varkappa}{6} u_* h \tag{2}$$

with the friction velocity $u_* = \bar{u}/c$. Equation (2) is only valid in cases where the flows are determined by the friction forces. If not, more accurate turbulence models like the k-ε model have to be used.

3 SEDIMENT MODEL

The mobile bed modelling follows the physical framework outlined by Spasojevic and Holly.[8,9] In this approach, a distinct separation of suspended and bedload transport for different grain size classes is made. The exchange between the two transport modes occurs in the mixing layer (i.e. bed surface layer) whose thickness depends on the composition of the bed material. The unknowns are the concentration c_k of the suspended grains, the fractional representation in the mixing layer β_k and the surface level of the bed z_b. The subscript k denotes the size class of the grains.

Assuming that the grains in suspension are transported downstream at the water velocity, the depth-averaged transport equation for the suspended sediment can be written as

$$\left(h c_k \right)_{,t} + \left(p c_k - \frac{h \nu_t}{\sigma_t} c_{k,x} \right)_{,x} + \left(q c_k - \frac{h \nu_t}{\sigma_t} c_{k,y} \right)_{,y} + s_k = 0 \tag{3}$$

with σ_t =turbulent Prandtl number (=0.5) and s_k =suspended load source. The other variables are defined according to Equations (1) and (2). The source term s_k describes the process of deposition and entrainment

$$s_k = w_k \left(c_{d_k} - \beta_k \, c_{e_k} \lambda_k \right)$$

where w_k is the fall velocity for particles of the size class k and c_{dk} is an empirical relation which describes the deposition near-bed concentration.[10] c_{ek} and λ_k denote the empirical entrainment near-bed concentration and the transport mode allocation parameter, respectively.[11] λ_k accounts for the fact that not all fractions are transported as suspended load but rather as bedload depending on the ratio of bed shear velocity to fall velocity.

The volume conservation in a mixing layer elemental volume for size class k can be written as

$$(1 - r)(\beta_k h_m)_{,t} + (p_{b_k})_{,x} + (q_{b_k})_{,y} - s_k + s_{m_k} = 0 \tag{4}$$

with r =porosity, h_m =thickness of the mixing layer according to Ref. 12 and (p_{b_k}, q_{b_k}) = components of bedload flux per unit width. The bedload flux for non-uniform sediments is evaluated as

$$p_{b_k} = (1 - \lambda_k)\beta_k\zeta_k \cdot p_{be_k}$$

where the equilibrium capacity p_{be_k} is determined with a bedload formula for uniform sediments (e.g. formula of Meyer–Peter). In the presence of larger grains the transport of finer fractions is reduced. According to Ref. 13, we obtain for the hiding factor the empirical relation

$$\zeta_k = \left(\frac{d_k}{d_{50}}\right)^{0.85}$$

where d_{50} and d_k denotes the diameter of the mean grain and the grain of fraction k, respectively. The exchange of sediment particles between the mixing layer and the underlying material with the size fraction β_{s_k} is expressed by the mixing layer source term

$$s_{m_k} = (1 - r)\big((z_b - h_m)\beta_k\big)_{,t}$$

if the mixing layer floor is rising and

$$s_{m_k} = (1 - r)\big((z_b - h_m)\beta_{s_k}\big)_{,t}$$

if the mixing layer floor is descending. Finally, the global volume conservation of bed material can be written as

$$(1 - r)z_{b,t} + \sum_{k=1}^{n}\Big(p_{b_k,x} + q_{b_k,y} + s_k\Big) = 0 \tag{5}$$

with n=total number of fractions.

4 DISCRETE SOLUTION

Equation (1) is solved on a standard grid with an implicit time stepping procedure. Implicit schemes, although more time consuming than explicit schemes, offer a better control of the convergence, especially for flow cases where drying and flooding occur. While the original Beam and Warming scheme is non-iterative, the strong non-linearities in the case of flooding and drying of grid cells suggest the use of an iterative procedure.

The mobile bed and the flow equations can be solved separately because the changes in bed level between two time steps are kept small. For the same reason the transport equations are solved one after the other leading to a considerable saving in CPU time.

4.1 Time differencing

In a semi–discrete form (1) can be written as

$$\frac{u^{n+1} - u^n}{\Delta t} = r(u) \tag{6}$$

where the superscripts n and $n+1$ denote old and new time level, respectively. All terms except the time derivation are included in the residuum r. The variable u can be expressed as a linear function of the old and the new (still unknown) values

$$u = \theta u^{n+1} + (1 - \theta)u^n \tag{7}$$

and the new value is then

$$u^{n+1} = u^n + \frac{u - u^n}{\theta} \tag{8}$$

The presence of the non-linear residuum vector precludes a direct solution of Equation (6). An iterative scheme is obtained using a local Taylor expansion about the former estimation of the unknowns u_o

$$r(u) = r(u_o) + r_{,u}\Delta u + O(\Delta u^2) \tag{9}$$

with the delta variable $\Delta u = u - u_o$ and the jacobian of the residuum $r_{,u}$. With Equations (8) and (9) in (6) a linear equation is obtained

$$\left(\frac{I}{\theta \Delta t} - r_{,u}\right)\Delta u = r(u_o) - \frac{u_o - u^n}{\theta \Delta t} \tag{10}$$

that can be solved with a factored algorithm.[6] The stability limits for the time factor are $0.5 \leq \theta \leq 1.0$. For $\theta = 1.0$ we obtain a fully-implicit backward difference procedure (Euler implicit) of first order accuracy. For $\theta = 0.5$ the scheme is equivalent to the trapezoidal rule in the linear case and the accuracy is approximately of second order. Convergence is achieved when the absolute value of the right-hand side of Equation (10) falls below a small given value.

4.2 Spatial discretization

The residuum in semi-discrete form is

$$r = -\frac{f_e - f_w}{\Delta x} - \frac{g_n - g_s}{\Delta y} - s^* \tag{11}$$

where the subscripts of the fluxes indicate the side of the cell (Figure 1).

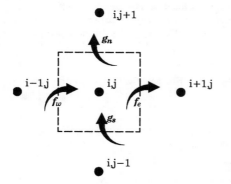

Figure 1
Discrete cell and nodes on a standard grid.

The discrete fluxes are now expressed by nodal values on a standard grid where all the flow parameters are located in the center of discrete cells. For the flux across the east side, for example, we obtain using central differences

$$
\boldsymbol{f}_e = \begin{bmatrix} \frac{1}{2}\left(p_{i,j} + p_{i+1,j}\right) \\[2mm] \frac{1}{2}\left(p_{i,j}u_{i,j} + p_{i+1,j}u_{i+1,j}\right) - 2\left(\frac{h\nu_t}{\Delta x}\right)_e \left(u_{i+1,j} - u_{i,j}\right) \\[2mm] \frac{1}{2}\left(p_{i,j}v_i + p_{i+1,j}v_{i+1,j}\right) - \left(\frac{h\nu_t}{\Delta x}\right)_e \left(v_{i+1,j} - v_{i,j}\right) - \left(\frac{\tau_{xy}}{\varrho}\right)_e \end{bmatrix}
$$

with the cross-derivative term explicitly treated as

$$
\left(\frac{\tau_{xy}}{\varrho}\right)_e = -\left(\frac{h\nu_t}{2\Delta y}\right)_e \left(u_{i,j}^n + u_{i+1,j+1}^n - u_{i,j-1}^n - u_{i+1,j}^n\right)
$$

and the discrete source term

$$
\boldsymbol{s}^* = \begin{bmatrix} 0 \\[2mm] gh_{i,j}\dfrac{z_{i+1,j} - z_{i-1,j}}{2\Delta x} + \left(\dfrac{\bar{u}u}{c^2}\right)_{i,j} \\[2mm] gh_{i,j}\dfrac{z_{i,j+1} - z_{i,j-1}}{2\Delta y} + \left(\dfrac{\bar{u}v}{c^2}\right)_{i,j} \end{bmatrix}
$$

In the transport equation of the suspended sediments (3), the fluxes are computed by a first-order upwind-weighted scheme providing a stable and low cost solution. The inherent numerical dissipation of this scheme does not affect the accuracy of the results because steep gradients in the variables hardly occur in natural water courses. This might be different in the case of suspended load releases from reservoir flushing operations.

4.3 Artificial dissipation

In the scheme described above, it is necessary to add dissipative terms to damp the short wavelengths. For this reason a fourth-order term is introduced that damps the oscillations but does not disrupt the formal accuracy of the scheme. The dissipation can be introduced as an additional flux. At the east side we obtain for the corrected flux

$$
\boldsymbol{f}_{ec} = \boldsymbol{f}_e - a_e\left(\boldsymbol{u}_{i+2,j} - 3\boldsymbol{u}_{i+1,j} + 3\boldsymbol{u}_{i,j} - \boldsymbol{u}_{i-1,j}\right) \tag{12}
$$

with the scaling velocity a_e depending on the local flow velocity and the wave velocity

$$
a_e = \varepsilon^{(4)}\left(|u| + \sqrt{gh}\right)_e \tag{13}
$$

and the coefficient $\varepsilon^{(4)}$ set to 0.1. An overview over various dissipation models including second-order terms is given in Ref. 14.

4.4 Dry cells

Often the internal boundaries are not known a priori or depend on the flow conditions. Because in dry cells the momentum equation is not valid, it is important that the internal boundaries are correctly treated. For the standard grid arrangement we can state the following rules:

i) A cell is dry if the flow depth is smaller than a given value h_{dry}.

ii) The flux across an internal boundary is zero if the water level in the wet cell is smaller than the bed level plus a given value h_{wet} in the dry cell .

To get a stable scheme the relation $h_{wet} \geqq h_{dry}$ must be fulfilled. Additionally the flux out of a dry cell is zero and the interpolation of the water level at internal boundaries must be adapted. Flooding and drying are non-linear processes by nature and may therefore disturb the stability of the numerical solution. As a consequence, the size of the time step has to be reduced to maintain convergence.

5 APPLICATIONS

The described numerical method is hardly applied in shallow water modelling. Thus, it is useful to test the scheme with some standard test cases as the dambreak on a flat plain. A more comprehensive analysis for the ability of the scheme can be found in Ref. 15. In a second, still one-dimensional example, the ability of flooding and drying is demonstrated. Last, the simulation of a flood event in an alpine valley shall emphasize the practical use of the algorithm and the need for mobile bed modelling.

5.1 Dambreak on a flat plain

The 1-d dambreak in a frictionless channel with flat bottom is considered. The initial flow depth upstream of the dam (at x=0m) is 1.00 meter. If the downstream region is initially dry, a simple wave is created as the dam is removed (Figure 2). A shock wave with an advancing discontinuity is formed if the downstream region is initially wet (Figure 3). Front position and shock height are well predicted by the numerical solution, whereas the discontinuities are smeared over several nodal values. Increasing the time accuracy can further improve the resolution of the discontinuities.

5.2 Dambreak on a slope

To illustrate the calculation of flooding and drying we consider the sudden breaking of a dam on a slope of 4% in a rectangular channel. Channel width and initial flow depth behind the dam are 0.30m, and the sand roughness is taken to be 1.5mm. The numerical results are compared with measured time series at fixed positions taken from Ref. 16. Figure 4 shows that the front positions are well captured by the calculation (dots), whereas the peak values of flow depth and discharge are slightly overestimated.

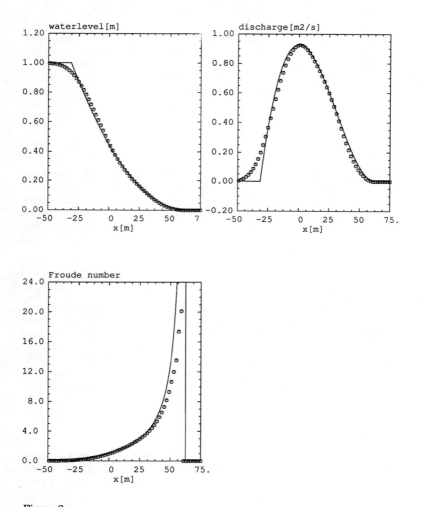

Figure 2
Dambreak on a dry plain.
Comparison between exact solution (solid line) and calculation after 10s.
(Δx=2.0m, Δt=0.5s, θ=1.0)

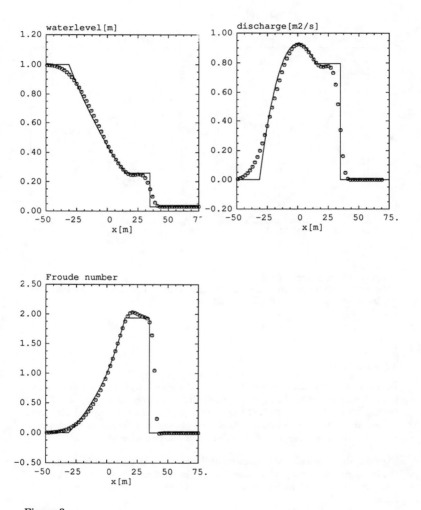

Figure 3
Dambreak on a plain with initial water level of 0.03m.
Comparison between exact solution (solid line) and calculation after 10s.
(Δx=2.0m, Δt=0.5s, θ=1.0)

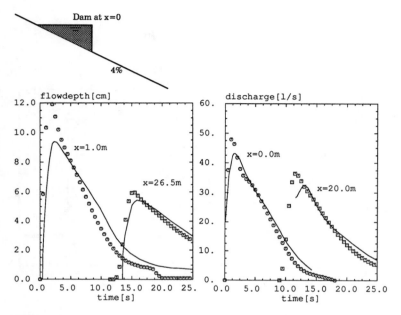

Figure 4
Dambreak on a slope of 4%.
Comparison between measured time series (solid line) and calculation.
(Δx=1.0m, Δt=0.25s, θ=1.0, k_s=1.5mm)

5.3 Flood in the Reuss valley

In order to test the model under real world conditions, a plain in the lower part of the Reuss valley has been simulated for the flood event in August 1987. After the break of the left levee of the Reuss river the plain was inundated. The inflow filled the area behind and as a consequence, the levee was overtopped with the water flowing back to the river forming two new breaches.[17]

Figure 5 shows calculated sequences of the flooding at different instants. For the computation, the area has been discretized with rectangles of 20x10m grid size. Four grain size fractions have been considered – two of them as suspended load.

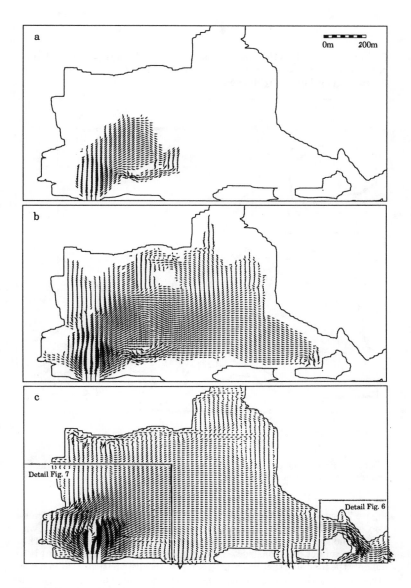

Figure 5
Simulated flooding of a plain in the lower part of the Reuss valley
at time (a) 0.5h (b) 0.7h and (c) 1.8h in August 25, 1987.

In Figure 6 a detail of the flow field is shown where transcritical conditions occur with Froude numbers up to 2.0. The stability of the scheme is not affected even for higher Froude numbers.

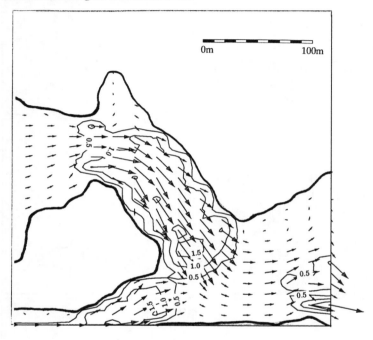

Figure 6
Detailed view of the calculated flooding and contour lines of the Froude number distribution at time 1.8h.

Figure 7 depicts a comparison between a clear water simulation and a calculation on mobile beds. The eroded material of the breach is deposited downstream and as a result the jet is widely spread to the east and west side. This behavior can not be reconstructed if only clear water is considered. But it is not only the flow field that looks different. Also the distribution of flooded and dry areas may change if the calculation considers mobile bed or not. In this case, the additional need on CPU time for mobile bed calculations does not exceed 20% of the total time consumption.

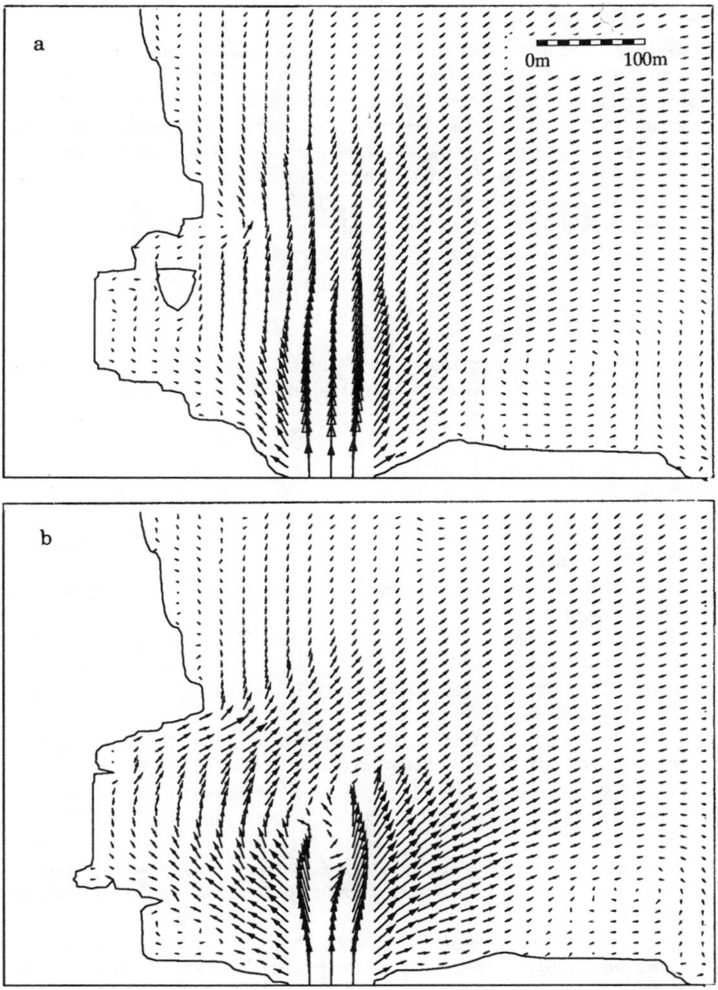

Figure 7
The flow pattern at the inflow breach at time 1.8h.
(a) clear water simulation (b) mobile bed simulation

6 CONCLUSIONS

Stability and robustness of an implicit, finite-volume scheme with central spatial differencing is demonstrated under mountainous flow conditions. Using a standard grid arrangement, the treatment of internal boundaries and small flow depths can be simplified. However, the algorithm can be further improved to reduce the time step limits and the CPU time needed. On the other hand, the additional expense for mobile bed simulations is small if the transport equations are separately solved. The applications show that, together with the mobile bed module, the described shallow water model represents a practical tool for surface flows under difficult topographic situations.

ACKNOWLEDGMENT

This work has been supported by *Landeshydrologie und Geologie, Bern,* in the frame of a research contract on the modelling of transport and mixing in rivers.

REFERENCES

[1] Cunge J.A., Holly F.M.Jr and Verwey A., *Practical Aspects of Computational River Hydraulics*, Pitman, 1980.

[2] Abbott M.B., *Computational Hydraulics: Elements of the Theory of Free Surface Flows*, Pitman, 1979.

[3] Abbott M.B. and Basco D.R., *Computational Fluid Dynamics: An Introduction for Engineers*, Longman, 1989.

[4] Glaister P., *Approximate Riemann solutions of the shallow water equations*, J.Hyd.Res., **26**, No 3, 293–306, 1988.

[5] Yang J.Y., Hsu C.A. and Chang S.H., *Computations of free surface flows, part 1: One-dimensional dam-break flow*, J. Hyd. Res., **31**, No 1, 19–34, 1993.

[6] Beam R.M. and Warming R.F, *An implicit finite-difference algorithm for hyperbolic systems in conservation-law form*, J. Computational Physics, **22**, 87–110, 1976.

[7] Bezzola G.R., *The effect of sediment transport during the 1987–flood in the Reuss River* , Proc. Int. Grain Sorting Seminar, 331–343, Monte Verità, Switzerland, 1991.

[8] Spasojevic M. and Holly F.M. Jr, *MOBED2–Numerical simulation of two-dimensional mobile bed processes*, IIHR Report No. 344, Iowa Inst. of Hydraulic Research, 1990.

[9] Holly F.M. Jr and Rahuel J., *New numerical / physical framework for mobile bed modelling, Part I – Numerical and physical principles*, J. Hyd. Res., **28**, No. 4, 401–416, 1990.

[10] Lin B., *Current Study of unsteady transport of sediment in China*, Proc. of Japan–China Bi–Lateral Seminar on River Hydraulics and Engineering Experiences, 23 Jul– 6 Aug, Tokyo–Kyoto–Sapporo, 1984.

[11] van Rijn L.C., *Sediment transport, part II: Suspended load transport*, J. Hyd. Eng., **110**, No. 11, 1984.

[12] Bennet J.P. and Nordin C.F., *Simulation of sediment transport and armoring*, Hydrological Science Bulletin, XXII, **4**, No 12, 1977.

[13] Karim M.F., Holly F.M.Jr. and Yang J.C., *IALLUVIAL, Numerical simulation of mobile bed rivers; part I, Theoretical and numerical principles*, IIHR Report No. 309, Iowa Inst. of Hydraulic Research, 1987.

[14] Pulliam T.H., *Artificial dissipation models for the Euler equations* , AIAA Journal, **24**, No. 12, 1931–1940, 1986.

[15] Beffa C., *Numerical methods on staggered and non-staggered grids for the shallow water equation*, submitted to Advances in Water Resources.

[16] Chervet A. et Dallèves P., *Calcul de l'onde de submersion consécutive à la rupture d'un barrage*, Schweizerische Bauzeitung, 88. Jg., Heft 19, 420–432, 1970.

[17] Faeh R., Koella E. and Naef F., *The flood in the Reuss valley in August 1987: A computer aided reconstruction of a flood in a mountainous region* , Proc. Int. Conf. on River Flood Hydraulics, Wallingford, England, 1990.

**Reliability and Validity of Modeling Sedimentation
and Debris Flow Hazards over Initially Dry Areas**

Douglas L. Hamilton, M.ASCE
Consulting Hydrologic Engineer
15991 Red Hill Avenue, Suite 200, Tustin, CA 92680

Robert C. MacArthur, M.ASCE
Principal, Northwest Hydraulic Consultants
1477 Drew Avenue, Suite 105, Davis, CA 95616

Vito A. Vanoni, F.ASCE
Professor Emeritus, California Institute of Technology
Keck Hydraulics Laboratory, Pasadena, CA 91125

ABSTRACT

Extreme floods may inundate populated areas of a river valley that are normally
dry. Two-dimensional flow analysis is a useful method for predicting the spatial
and temporal behavior of water as it leaves a defined channel and enters an
initially dry area. Although the analysis of flow behavior in regions of
discontinuous or unbounded geometry is complex, the governing equations are
known and generally agreed upon. When the water contains a significant amount
of sediment or the unbounded region is subject to scour and deposition, the flow
behavior becomes less predictable. Results from both clear water, fixed boundary
simulation models and models that attempt to incorporate sedimentation effects
must be carefully evaluated to determine their validity. The purpose of this paper
is to discuss the role of reliability and validity in analyzing the potential
sedimentation and flood hazards in off-channel areas.

Watersheds that yield large amounts of granular sediment often create depositional
fans in the downstream regions. Over time, as these regions become densely
populated, flood control facilities are designed and built. Such facilities may fail
or overtop when the combined discharge of water and sediment exceeds the
capacity of the system. Failure may also occur if bridge crossings become
blocked with debris or if a channel experiences extreme sediment deposition.

In order to assess the validity of computed results, two factors must be considered in the analysis of sedimentation/flooding events. First, the presence of sediment may effect the characteristics of the fluid itself rendering traditional resistance formulas inapplicable. This is the case for debris flows which are often governed by laminar resistance when they deposit over initially dry areas. Second, the presence of sediment or debris within the source water may drastically impact the characteristics of the receiving area thus blocking off preferential flow paths and flooding areas that may have been previously considered *safe*.

WHAT DO RELIABILITY AND VALIDITY MEAN?

Definition
The primary purpose of this paper is to present a framework in which one can fully appreciate the results yielded from analytical methods that are applied to extremely complex physical processes. With this in mind the following definitions of reliability and validity are offered.

> **Reliability** is the quality of a method yielding similar results for successive applications which are implemented by different people. Furthermore, reliable methods have the essential characteristic of confident use based on a tradition of successful applications.

> **Validity** is the quality of a result being at once relevant, meaningful, and logically correct. A result is relevant if it addresses the significant issues of the matter at hand. It is meaningful if it conveys information that can be the basis for specific actions or decisions. A result is logically correct if it is within ranges established by observed data, historical experience, and physical limitations.

Reliability applies to methods, models, or techniques. Validity applies to the results obtained from the application of such methods, models, or techniques.

Importance
Methods are considered reliable when they consistently yield valid results. The criteria for determining the validity of results are sometimes difficult to define for complex flooding processes. It is possible, therefore, that due to a lack of discernment about what a valid result should look like, one may erroneously conclude that a particular method is either reliable or unreliable.

The authors' experience indicates that when analyzing the potential hazard caused by a complex flooding situation, it is important to include an assessment of how adequately a proposed method incorporates the significant physical processes that govern a particular application. Furthermore, it is often advantageous to draw upon the expertise of a wide range of professionals both in the evaluation of an existing flood hazard and the appropriate means for its remediation.

EFFECTS OF SEDIMENT ON THE BEHAVIOR OF THE FLUID

High concentrations of sediment and debris in flowing water can cause it to behave differently than clear water flows. Some of these differences, such as the unit weight, are quantitative in nature. Other differences, such as the vertical velocity distribution for a debris flow, display qualitative differences compared to clear water. These differences and their effect on the behavior of the fluid are discussed here.

Classification
The literature contains several classification procedures for distinguishing between the different type of sedimentation hazards. Most of these use the volumetric concentration of sediment in the fluid (Kurdin, 1973; Campbell, 1985). Other procedures such as described in Miranova and Yablonskiy (1992) are based on the incipient cause of the flow such as a landslide, a volcanic eruption, a dam failure, etc. For flood investigations, the primary events of concern can be generally grouped into the following three categories: clear-water flows which can be analyzed with traditional hydraulic methods, hyperconcentrated sediment flows which can be analyzed to a great extent by sediment transport theory, and debris flows which can be assessed by various empirical methods such as the bulking factor, the Bingham model, etc.

Vertical Velocity Distribution
For turbulent clear water flows, the vertical velocity distribution is essentially a function of the logarithm of the ratio of the distance above the bed to the total depth of flow (Rouse, 1936). Vanoni (1953) suggested that as the suspended sediment concentration increased, the friction factor f decreased in association with a decrease in the von Karman coefficient. Figure 1 shows the comparison of two measured velocity profiles. The one on the left is for clear water. The one on the right is for a flow with a suspended sediment concentration of 15.8 g/l. The depth averaged velocity for the clear water flow is about 70 cm/s. Under the same depth and slope, the flow with high suspended sediment load has a depth averaged velocity of approximately 93 cm/s. Although their is some debate as to the actual cause, it is important to recognize that high sediment concentrations can result in flows with a higher average velocity compared to clear water.

Takahashi (1978, 1980) developed equations of motion for a debris flow. He determined the vertical velocity distribution obeys the relationship:

$$u = K[h^{3/2} - (h-y)^{3/2}] \tag{1}$$

where u is the velocity at a distance y above the bed, h is the total depth of flow, and K is a lumped parameter that describe site specific geometric and fluid properties. Flume experiments were conducted that indicated this velocity distribution had better agreement than that of typical laminar Newtonian flow.

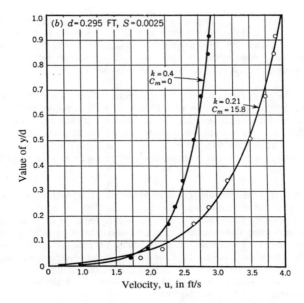

Figure 6.   Vertical velocity distributions from Vanoni's (1953) flume studies.

Zhang and Ren (1982) conducted similar flume experiments using higher sediment concentrations. Their studies focus on viscous flows with low Reynolds numbers. The experiments indicate that as suspended sediment concentration increases, the top layer of the flow behaves as a rigid plug flowing over a relatively narrow shearing zone that is closer to the bed. Many of these experiments were for flows in the laminar regime. The existence of a flowing plug on the top layer would explain the observation of non-rotating boulders that apparently *float* on the surface of a debris flow. Higher velocities compared to equivalent clear water flows were also reported.

Umeyama and Gerritsen (1992) present a theoretical model for the velocity distribution in sediment laden flow assuming that the von Karman coefficient remains at its traditional value of 0.4. The work suggests a modified method of determining the Prandtl mixing length can better explain the results of flume experiments. The research concluded that mixing of water and suspended sediments in the turbulent boundary layer is less effective. The meaning of this conclusion (as with the earlier work) is that, up to a point, higher sediment concentrations result in faster moving flow.

Rickenmann (1991) performed a series of flume studies using clay suspensions in a recirculating flume. Although his work did not focus on the determination of the velocity distribution, he concluded that an increase in bulk fluid density increased bed load sediment transport rate. The bed load transport rate began to decrease once the clay content of the suspension exceeded 17%.

The laboratory flume studies of Lyn (1991, 1992) conclude that examination of the velocity profile alone is not sufficient to conclude that overall flow resistance is always increased or decreased in comparison to the equivalent clear water flow. In fact, near-bed velocities in many sediment laden flow experiments were lower than for equivalent clear water flows. In these cases the velocity in the upper layers of flow as well as the depth averaged velocity was always higher than clear water flow.

Even though there remains debate as to the exact relationship between the friction factor and the vertical velocity distribution, there are two points of interest for use when modeling flows in one or two dimensions. First, the depth averaged velocity increases when flows contains high concentrations of suspended sediment load. Even though it may not be entirely correct to attribute this increase to a reduction in flow resistance, the net effect of this for most engineering or risk assessment purposes is the same as if it was entirely correct. Second, when the suspended load concentration is below the 15% to 20% range, the higher velocity results in an increased bed load transport capacity.

Sediment Transport Characteristics
The presence of high sediment concentrations in flood waters can have other effects besides the increased transport of bed load sediments. Bagnold (1954) recounts a *river of gravel* flowing through a mountain canyon with no apparent interstitial fluid. He explains it using the concept of dispersive stress which results from the collision of individuals particles in the flow. These collisions result in the larger particles being transported on the top of the flow rather than rolling on the bed.

Wang and Zhang (1990) investigated the interaction of clay suspensions flowing over a gravel bed. Although their investigation was primarily concerned with the behavior of debris flows, they arrived at several conclusions related to the types of sediment transport that may occur. The front of a debris flow often travels as a bore or shock and often carries large boulders. The trunk zone behind the bore carries coarse sediments (that are in suspension due to inter-granular collision) at a higher velocity than the bore itself. The coarse material rolls over the debris flow front continuously providing energy to it. This is one explanation for the presence of coarser material at the front of a debris flow.

Li, et al. (1983) describes field investigations of debris flows in the Jiang-Jia Ravine in Yunnan, China. On a field scale, the process described above results in a series of waves or surges each transporting large amounts of coarse sediment. Boulders larger than 6 meters in diameter have been carried downstream by this process. Wang and Zhang (1990) present a reason for the sorting process which results in the coarser material residing on the surface of the debris flow front. Resistance to the motion of a given particle in a debris flow is primarily due to collisions with materials on the bed. When a large particle collides with smaller bed material particles, its forward momentum remains relatively unchanged. On the other hand, when a smaller particles collides with materials on the bed it may completely lose its forward momentum. Large particles, therefore, have a much higher tractive force than resistive force and therefore tend to move faster, concentrating at the front of the debris flow.

Colby (1964) presents a procedure for estimating the effect of high suspended load concentrations on the bed load transport rate. His work is based on the Einstein bed load function and various sources of measured sediment transport data from both natural streams and laboratory flumes.

One can see that the sediment transport potential and flow characteristics of a flood event can be significantly affected by changes in the bulk fluid properties. Traditional bed load transport equations and hydraulic calculations may need to be supplemented with other methods in order to adequately describe the flow behavior.

Turbulent v. Laminar Regime
Most analysis of flooding events relies on theory developed for full turbulent flows. In these cases, resistance to flow is determined primarily by the roughness characteristics of the flow surface. Flows with high concentrations of sediment, however, can often be dominated by viscous rather than inertial forces. In this case the properties of the sediment-water mixture deviate significantly from clear water. The fluid may possess both a finite yield strength and a non-linear relationship between the shear stress and shearing rate. Bingham (1922) proposed the following model:

$$\tau = \tau_y + \eta \left(\frac{du}{dy}\right) \qquad (2)$$

where τ is the shear stress in the fluid; τ_y is the Bingham yield strength; η is the viscosity; and du/dy is the velocity gradient in the primary flow direction. More complex models have been developed by Chen (1988) and Julien and Lan (1991) which also consider dispersive and turbulent stresses. These models have the general form:

$$\tau = \sum_{j=0}^{n} C_j \left(\frac{du}{dy}\right)^j \qquad (3)$$

The first term of this equation C_0 is the Bingham yield strength τ_y. The second term C_1 is the fluid viscosity η. The third term C_2 is the dispersive / turbulent stress coefficient. The fourth term C_3 is the shear thinning coefficient, etc. In order to apply these concepts to hydraulic computations, Equation (3) can be can be rearranged to a flow resistance equation of the form:

$$S_f = \frac{M\tau_y D}{64\gamma yR} + \frac{M\eta D^2}{32\gamma y^2 R^2}V + \frac{n^2 D^{4/3}}{2.22y^{4/3}R^{4/3}}V^2 \qquad (4)$$

Where S_f is the energy grade line slope; D is the hydraulic depth; γ is the composite unit weight; R is the hydraulic radius; M is the laminar roughness coefficient (as in $f=M/Re$); and n is Manning's roughness coefficient.

Zhang S. (1992) indicates that the potentially most destructive sediment flow event is the viscous debris flow. A schematic representation of this type of event is shown in Figure 2. Levees often form on the perimeter of the debris flow which help to channelize it (Figure 2a.). Deposition lobes often form at the end of these channels. Slopes in excess of 4% may result in flow surges as the debris travels downstream (Figure 2b) while milder slopes result in debris flows of uniform thickness (Figure 2c). Deposits are sometimes left along the edges of a debris flow (Figure 2d). Large boulders can be supported and carried by the frontal surge of the flow (Figure 2e). Figure 3 shows a post-event photograph of a viscous debris flow in gravelly material.

Assuming the top layer of a viscous debris flow behaves as a rigid plug, Zhang S. developed a method to estimate the potential flow velocity (Hamilton, et al., 1993). This method uses a revised version of Equation 2:

$$\frac{u_{max}-u}{u_{max}} = (1 - \frac{\tau}{\tau_0 - \tau_y})^2 \qquad (5)$$

Where u_{max} is the velocity of the plug layer; and τ_0 is the shear stress on the stream bed. Integrating this expression gives:

$$V = \frac{\tau_0 h}{3\eta}[1 - \frac{3}{2}(\frac{\tau_y}{\tau_0}) + \frac{1}{2}(\frac{\tau_y}{\tau_0})^3] \qquad (6)$$

By noting the $\tau_0 = \gamma RS_f$ it can be shown the mean velocity of a one dimensional debris flow is given by:

$$V = \frac{R}{\eta}(\frac{32\gamma RS_f}{M} - \frac{\tau_y}{2}) \qquad (7)$$

The above equation is valid for viscous debris flows in the laminar regime.

EFFECTS OF SEDIMENT ON THE INITIALLY DRY AREA.

Sediment transport processes can drastically change the behavior of a flood both in a defined channel and after it leaves the channel. During an event, the erosion process may have the effect of augmenting or *bulking* the hydrograph so that peak flows are significantly larger than those estimated from rainfall intensities. Furthermore, the sediment deposition process may alter the geometry that governs the flow depth, velocity, and direction in the initially dry area.

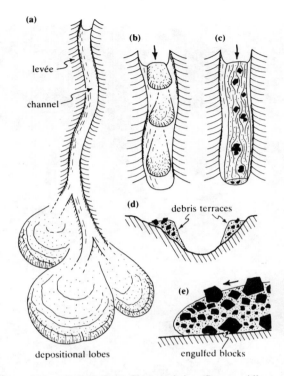

Figure 2.   Debris Flow Characteristics (Source: Allen, J.R.L., 1985, Principles of Physical Sedimentology, Chapman and Hall, London.

Erosion Process

Blackwelder (1928) describes how steep channels in semi-arid mountains create ideal mixing troughs for water and sediment. During a heavy rainfall season the sides of steep canyons become saturated, fail, and slide into the stream channel. Runoff from subsequent events picks up this loose, saturated material and conveys it downstream. The increased bulk density creates favorable conditions for picking up additional sediment stored in the stream bed. The result is a debris flow that has a larger volume and peak flow than one can attribute solely to runoff from its coincident rainfall event. Examples of similar occurrences are discussed in Marsell (1970), Slosson (1986), and Anderson (1984).

Figure 3.   Post-event photograph of a debris flow in a gravelly material.

Tagomori (1988) describes the New Year's Eve 1987 flood in the eastern part of Oahu, Hawaii. A storm dropped as much as 60 cm of rain in a 24 hour period with intensities exceeding 10 cm/hr. The Hahaione Valley channel eroded wider and deeper to such an extent that an existing settling basin was quickly filled, blocking a culvert entrance. Flows left the constructed channel and were conveyed through downhill sloping streets. Shearing forces created by the diverted flows were so great that they ripped up slabs of asphalt pavement and eroded a channel approximately 2 meters deep and 5 meters wide inside the street alignment. Such changes in geometry are difficult to predict from a computational standpoint. They do, however, have a significant effect on the flood behavior in the previously dry area.

MacArthur et al. (1991) indicate that the peak flows of the 1987/88 Oahu event described above were initially characterized as having 500-year recurrence interval. Landslides in the upper watershed and saturated soil scoured from the channel bed contributed to the severity of these flows. Absent this bulking effect, it was later determined that the flows attributable to direct runoff from rainfall had a recurrence interval of 20 to 50 years. The point of this narrative is that sedimentation processes may affect the design hydrograph which is an essential boundary condition for many flow modeling efforts.

The bulking factor is a common means of incorporating the potential effects of erosion upon escalating the design hydrograph. For example, assuming that the maximum concentration by volume, C_v, of a debris flow is 0.5, the runoff attributable to rainfall may be multiplied by a factor of $1/(1-C_v)$ or 2 to come up with a hydrograph that represents the effects of a debris flow. This approach is commonly used as a safety factor for hydraulic calculations used done for the design of flood control facilities. Quite unfortunately its misuse is also becoming common.

Volcanic deposits, burned watersheds, and landslide blockages can create the potential for extreme flow events that cannot be characterized by the simple assessment of maximum concentration (Guy, 1970; Hamilton, et al., 1986; MacArthur, et al., 1990; Pierson, et al., 1992; and Slosson, et al., 1986). Erosional processes often mobilize additional water that saturates the pores of underlying strata. Schaefer (1992) demonstrates that for certain events, the *ultimate* bulking factor can range from 1.1 to more than 10.0. He developed the following relationship:

$$BF_{ult} = \frac{1+\phi}{1-\phi\theta} \qquad (8)$$

where ϕ is the ratio $1/(1-C_v)$ with C_v being the maximum expected concentration by volume of the event; and θ is the effective void ratio (the product of the void ratio and the fraction of voids filled with water). Equation (8) was used to estimate a bulking factor of 3.32 for peak flows from a potential failure of a landslide lake blockage near Mount St. Helens in the U.S. (MacArthur, et al., 1990).

On the other side of the spectrum, the bulking factor is sometimes incorrectly used as a calibration factor for bed and suspended load calculations. Since its primary purpose is to introduce a factor of safety into the hydraulic design of a facility, it will not necessarily match with observed or computed sediment transport data. Furthermore it may distract one from the more significant sedimentation issues at hand (Saarinen et al., 1984).

Deposition Process
When considering the impacts of floods on initially dry areas, the process of sediment deposition can significantly influence the direction and extent of flood flows. The alluvial fan, which is itself the product of deposition, illustrates this point. Often, the existing channel or preferential flow path is under capacity compared to the design event. When the design event occurs, it is difficult to

determine in advance whether or not the flow will proceed along its historical course or seek a new direction.

Dawdy (1979) and Mifflin (1990) suggest that the risk potential from flooding on alluvial fans is best determined by relatively straightforward probabilistic methods. The National Flood Insurance Program of the United States accedes this perspective and prohibits the use of two-dimensional flow modeling for determining flow behavior on alluvial fans (Federal Emergency Management Agency, 1991). The Hydrologic Engineering Center (1993) outlines case studies of structural flood control measures that have been built to protect alluvial fan areas. The report confirms the fact that deposition of sediment during a flood event was often unanticipated in the design process and sometimes resulted in at least as much damage as the effects of erosion.

CLOSURE

Figure 4 shows a complex alluvial system. The alluvial fan shown in the center of the photograph emanates from a highly disturbed tectonic pressure ridge. This ridge line, which goes diagonally across the photograph, collects and concentrates runoff that comes from a dendritic channel system on the apron of a much larger alluvial fan upstream. Even though the formation of the alluvial fan was probably the result of a single convulsive event such as a breach of the ridge, hazards from more frequent floods still exist.

The prior discussion should indicate to the reader that application of clear-water, fixed boundary hydraulic models to the situation shown in Figure 4, whether one- or two-dimensional, may not adequately characterize the overall flood behavior. Developing more complex models that attempt to quantify the effects of sedimentation may not necessarily be the complete solution to our dilemma, however. The reason is that in the former case we know the methods are somewhat unreliable considering their neglect of sedimentation processes. In a sense, we know the answer is not completely right. In the latter case we may develop confidence by implementing the more complex methods but the validity of the results may come into question due to the lack of observed data with which to compare it. In one case, we know that we are wrong. In the other case, we are not sure that we are right.

Modeling alone will not achieve the reliability and validity in studies of complex flooding conditions. When modeling applications are combined with an adequate appreciation for the importance of both hydraulic and sedimentation processes, however, these goals are within reach.

Figure 4.   Photograph of a complex alluvial system.

REFERENCES

Anderson, L.R., Keaton, J.R., Saarinen, T.F., and Wells, W.G., 1984, The Utah Landslides, Debris Flows, and Floods of May and June 1983, National Academy Press, Washington, D.C.

Bagnold, R.A., 1954, Experiments on a Gravity-Free Dispersion of Large Solid Spheres in a Newtonian Fluid Under Shear, Proceedings of the Royal Society of London.

Bingham, E.C., 1922, Fluidity and Plasticity, McGraw-Hill, New York, NY.

Blackwelder, E., 1927, Mudflow as a Geologic Agent in SemiArid Mountains, Bulletin of the Geological Society of America, V. 39, 1928, June.

Campbell, R.H., 1985, Landslide Classification for Identification of Mudflows and other Landslides, Open File Report 85-276, U.S. Geological Survey, Reston, Virginia, 1985.

Chen, C.L., 1985, Hydraulic Concepts in Debris Flow Simulation, in Delineation of Landslide, Flash Flood, and Debris Flow Hazards in Utah, Publication No. UWRL/G-85/03, Utah Water Research Laboratory, Utah State University, Logan, UT, August.

Colby, B.R., 1964, Practical Computations of Bed Material Discharge, Journal of the Hydraulic Division, ASCE, v. 90, no. HY2.

Dawdy, D., 1979, Flood Frequency Estimates on Alluvial Fans, Journal of the Hydraulics Division, V.105, No. HY11, 1979, November.

Federal Emergency Management Agency, 1991, Flood Insurance Study Guidelines and Specifications for Study Contractors, FEMA 37, Washington, D.C.

Guy, H.P., Flood Flow Downstream from Slide, 1970, Journal of the Hydraulics Division, V. 97, No. HY4, ASCE, 1971, April.

Hamilton, D., Zhang, S., and MacArthur, R., 1993, Debris Flow Velocity Estimation Methods for Natural Hazard Assessment, Proceedings of the ASCE 1993 National Conference on Hydraulic Engineering, San Francisco, California.

Hamilton, D., 1990, Mitigation of Sedimentation Problems in Watersheds at the Urban Interface, Proceedings of the California Watersheds at the Urban Interface, Watershed Management Council, Ontario, California.

Hydrologic Engineering Center, 1993, Assessment of Structural Flood Control Measures on Alluvial Fans, U.S. Army Corps of Engineers, Davis, CA.

Julien, P.Y., and Lan, Y., 1991, Rheology of Hyperconcentrations, Journal of the Hydraulics Division, ASCE, v. 117, no. 3.

Kurdin, R.D., 1973, Classification of Mudflows, Collection of Soviet Papers on Hydrology, No. 11, Issue No. 4.

Li, J., Jianmo, Y., Cheng, B., and Liu, D., 1983, The Main Features of the Mudflow in Jiang-Jia Ravine, Journal of Geomorphology, V. 27, No. 3, Berlin-Stuttgart, Germany, 1983, September.

Lyn, D.A., 1991, Resistance in Flat-Bed Sediment-Laden Flows, Journal of Hydraulic Engineering, V. 117, No. 1, ASCE, January.

Lyn, D.A., 1992, Turbulence Characteristics of Sediment Laden Flows in Open Channels, Journal of Hydraulic Engineering, V. 118, No. 7, ASCE, July.

MacArthur, R.C., Harvey, M.D., MacArthur, T.B., Hamilton, D.L., and Kekaula, H., 1991, Urban Flooding and Debris Flow Analysis for Niu, Aina Haina, and Kuliouou Valleys, A Reconnaissance Level Report Prepared for the U.S. Army Corps. of Engineers, Contract No. DACW83-91-P0055.

MacArthur, R., Brunner, G., and Hamilton, D., 1990, Numerical Simulation of Mudflows from Hypothetical Failures of the Castle Lake Debris Blockage Near Mount St. Helens, WA.,: Final Project Report No. 90-05, Hydrologic Engineering Center, Davis, California.

MacArthur, R., Schamber, D., and Hamilton, D., 1988, Incorporating the effects of mudflows into flood studies on alluvial fans: Special Projects Report No. 86-4, Hydrologic Engineering Center, Davis, CA.

Marsell, R.E., 1971, Cloudburst and Snowmelt Floods, Environmental Geology of the Wasatch Front, 1971: Utah Geological Association Publication 1-N., Salt Lake City, Utah.

Mifflin, E., 1990, Entrenched Channels on Alluvial Fans, Proceedings of The ASCE International Symposium of Hydraulics/Hydrology of Arid Lands.

Miranova E.M., and Yablonskiy, V.V., 1992, A Mathematical Model of Shear Debris Flow, Erosion, Debris Flows, and Environment in Mountain Regions: Proceedings of the Chengdu Symposium, IAHS Publ. No. 209.

Pierson, T.C., Janda, R.J., Umbal, J.V., Daag, A.S., Immediate and Long-Term Hazards from Lahars and Excess Sedimentation in Rivers Draining Mt. Pinatubo, Philippines: U.S.G.S. Water-Resources Investigations Report 92-4039, 1992, Vancouver, Washington.

Rickenmann, D., 1990, Hyperconcentrated Flow and Sediment Transport at Steep Slopes, Journal of Hydraulic Engineering, V. 117, No. 11, ASCE, 1991, November.

Rouse, H., 1936, Modern Conceptions of the Mechanics of Fluids Turbulence, Proceedings of the American Society of Civil Engineers, Paper No. 1965.

Saarinen, T.F., Baker, V.R., Durrenberger, R., and Maddock, T., 1984, The Tucson, Arizona, Flood of October 1983, National Academy Press, Washington, D.C.

Schaefer, A.D., 1992, Ultimate Bulking Factor for Mudflows, Conference on Steep Channels, Flood Control Channel Research Group, U.S. Army Corps of Engineers, Portland, Oregon.

Slosson, J.E., Shuirman, G., and Yoakum, D., 1986, Responsibility/Liability Related to Mudflow/Debris Flows, 1986, Proceedings of ASCE Water Forum, 1986, Long Beach, California.

Tagomori, M., Post Flood Report New Year's Eve Storm December 31, 1987-January 1, 1988 Windward and Leeward East Oahu: Circular C119, Department of Land and Natural Resources, Division of Water and Land Development. Honolulu, Hawaii, 1988, July.

Takahashi, T., 1978, Mechanical Characteristics of Debris Flow, Journal of the Hydraulics Division, V. 104, No. HY8, ASCE, 1978, August.

Takahashi, T., 1980, Debris Flow on Prismatic Open Channel, Journal of the Hydraulics Division, V. 106, No. HY3, ASCE, 1980, March.

Umeyama, M., and Gerritsen, F., 1990, Velocity Distribution in Uniform Sediment Laden Flow, Journal of Hydraulic Engineering, V. 118, No. 2, ASCE, 1992, February.

Vanoni, V.A., 1953, Some Effects of Suspended Sediment on Flow Characteristics, Proceedings of the Fifth Hydraulic Conference, Bulletin 34, Iowa State University, Studies in Engineering, Iowa City, Iowa.

Wang Z., and Zhang X., 1990, Initiation and Laws of Motion of Debris Flow, Proceedings of The ASCE International Symposium of Hydraulics/Hydrology of Arid Lands.

Zhang, R. and Ren, X., 1982, Experiments in Hyperconcentrated Flows, China Science, Beijing, China, August (in Chinese).

Zhang, S., 1992, A Comprehensive Approach to the Observation and Prevention of Debris Flows in China, Journal of Natural Hazards, August.

A comparison between computed and measured bed evolution in a river bend.

L. Montefusco, A. Valiani
Dipartimento di Ingegneria Civile
Universita' degli Studi di Firenze
via S. Marta 3
50139 Firenze ITALY

ABSTRACT

A shallow-water scheme with a movable bed has been employed in order to check the reliability of computations related to the bed evolution when a substantially three-dimensional phenomenon is present. The complete research program includes several laboratory experimental tests, as published data are generally not sufficiently complete to result suitable for a detailed comparison. Here only a first stage of the research is presented. In this paper a bend is considered, and shallow-water equations are solved in a polar coordinate system (the scheme is two-dimensional). A partially new numerical finite-difference technique has been developed, which seems to be specially robust, avoiding a staggered grid and allowing very long simulations. The results seems to be rather discouraging, as the scheme (at least in its more elementary version) seems not to be able to represent, even qualitatively, the experimental bed evolution.

INTRODUCTION

The availability of rather quick and cheap computing facilities and of sufficiently robust numerical techniques diffused in recent times the shallow-water scheme as a good tool for treating several free-surface hydraulic problems. Many of such problems are posed by the fluvial modelling, and can generally be said that the shallow-water scheme is very suitable for the evaluation of the local free-surface elevation and even for a general simulation of the flow patterns. Anyway one of the most interesting application of the scheme in a fluvial environment is obviously the apparent possibility to compute the local sediment transport and the local river bed evolution. This possibility is also considered in several commercial modelling packages.
It is well known that a fluvial flow presents very often three-dimensional flow patterns, and that the sediment erosion and deposition phenomena are very sensitive to secondary currents, which are not considered by the shallow-water scheme. It is certainly possible, and it has been done, to introduce some specific corrections to the scheme in order to simulate

effectively specific cases [ref. 1 to 15; look at 14 for a larger amount of references], but a test of the scheme, in its general form, for some common "complex" river conditions seems to be useful. In the present case a regular laboratory bend has been considered, with rigid vertical walls; the bed is supposed to consist of incoherent monogranular sediments. This choice has been done to avoid any spurious phenomena, due to an insufficiently accurate and smooth description of the flow-field geometry, which could superimpose effects that are not strictly due to the mathematical scheme. Further, this case has been widely studied and even comparisons between results of mathematical models and experiments exist [ref. 1 to 15].

However the previously employed mathematical models are different or in some sense more rough than the complete shallow-water scheme (adopting, for instance, a rigid lid approximation for the free surface) and the present simulation uses a more refined grid than usually done. Starting from a rest water condition and from an horizontal flat bed, a time marching simulation has been conducted for eleven hours and the results have been analysed, and qualitatively compared with the results of a similar experimental layout [5, 12].

THE MATHEMATICAL SCHEME

Assuming a polar r, θ coordinate system the momentum (along the θ and r directions) and mass balance equations for the liquid phase can be written as follows:

$$\frac{\partial q_\theta}{\partial t} + \frac{1}{r}\frac{\partial}{\partial \theta}\left(\frac{q_\theta^2}{y} + \frac{1}{2}\,g\,y^2\right) + \frac{\partial}{\partial r}\left(\frac{q_r q_\theta}{y}\right) + \frac{2}{r}\,\frac{q_r q_\theta}{y}$$

$$+ g\,y\,\frac{1}{r}\frac{\partial z_f}{\partial \theta} + \frac{1}{\rho}\,\tau_{b\theta} = 0 \qquad (1)$$

$$\frac{\partial q_r}{\partial t} + \frac{1}{r}\frac{\partial}{\partial \theta}\left(\frac{q_r q_\theta}{y}\right) + \frac{\partial}{\partial r}\left(\frac{q_r^2}{y} + \frac{1}{2}\,g\,y^2\right) + \frac{1}{r}\,\frac{q_r^2}{y}$$

$$- \frac{1}{r}\,\frac{q_\theta^2}{y} + g\,y\,\frac{\partial z_f}{\partial r} + \frac{1}{\rho}\,\tau_{br} = 0 \qquad (2)$$

$$\frac{\partial y}{\partial t} + \frac{1}{r}\frac{\partial q_\theta}{\partial \theta} + \frac{\partial q_r}{\partial r} + \frac{1}{r}\,q_r = 0 \qquad (3)$$

where:

$y(r,\theta,t)$ is the local instantaneous depth of the flow,

$q_\theta(r,\theta,t) = U_\theta\,y$ is the unit-width discharge in the θ direction,

$q_r(r,\theta,t) = U_r\,y$ is the unit-width discharge in the r direction,

$z_f(r,\theta,t)$ is the local instantaneous bed elevation,

g is the gravity acceleration,

ρ is the fluid density,

$\tau_{b\theta}$ and τ_{br} are the local instantaneous components of the bed shear stress.

As can be observed, no dispersive terms are here taken into account, because no general agreement exists for their representation and their correct expressions strongly depends on the secondary currents. The search for the possibility of a general effective expression of the dispersive terms is one of the future aims of the study. In this stage we have focused our attention on the propagation and the convection terms, which are clearly posed from a physical point of view, and which are dominants on the general flow behaviour.

The terms $\tau_{b\theta}$ and τ_{br} have been expressed as usually in the form:

$$\tau_{b\theta} = \rho \; q_\theta \; \frac{\sqrt{(q_\theta^2 + q_r^2)}}{Ch^2 \, y^2} \qquad\qquad \tau_{b\theta} = \rho \; q_r \; \frac{\sqrt{(q_\theta^2 + q_r^2)}}{Ch^2 \, y^2}$$

where Ch is the non-dimensional Chezy coefficient.

The mass balance equation for the solid phase is:

$$\frac{\partial z_f}{\partial t} + \frac{1}{r} \frac{\partial q_{s\theta}}{\partial \theta} + \frac{\partial q_{sr}}{\partial r} + \frac{1}{r} \, q_{sr} = 0 \tag{4}$$

where $q_{s\theta}$ and q_{sr} are the components of the unit-width solid discharge.

SOLUTION TECHNIQUE

The adopted solution is of the uncoupled type, having solved for each time step first the flow equations, and subsequently the bed evolution equation (4).

For the flow equations a classical ADI splitting technique has been used, considering for the θ direction the equations:

$$\frac{\partial q_\theta}{\partial t} + \frac{1}{r} \frac{\partial}{\partial \theta} \left(\frac{q_\theta^2}{y} + \frac{1}{2} g \, y^2 \right) + \frac{\partial}{\partial r} \left(\frac{q_r q_\theta}{y} \right) + \frac{2}{r} \; \frac{q_r q_\theta}{y}$$
$$+ g \, y \, \frac{1}{r} \frac{\partial z_f}{\partial \theta} + \frac{1}{\rho} \, \tau_{b\theta} = 0 \tag{1}$$

$$\frac{\partial y}{\partial t} + \frac{1}{r} \frac{\partial q_\theta}{\partial \theta} + \alpha \, \frac{1}{r} \, q_r = 0 \tag{5}$$

and for the r direction the equations:

$$\frac{\partial q_r}{\partial t} + \frac{1}{r} \frac{\partial}{\partial \theta} \left(\frac{q_r q_\theta}{y} \right) + \frac{\partial}{\partial r} \left(\frac{q_r^2}{y} + \frac{1}{2} g \, y^2 \right) + \frac{1}{r} \; \frac{q_r^2}{y} - \frac{1}{r} \; \frac{q_\theta^2}{y}$$
$$+ g \, y \, \frac{\partial z_f}{\partial r} + \frac{1}{\rho} \, \tau_{br} = 0 \tag{2}$$

$$\frac{\partial y}{\partial t} + \frac{\partial q_r}{\partial r} + (1 - \alpha) \, \frac{1}{r} \, q_r = 0 \tag{6}$$

where α is a constant with a value between 0 and 1 (its value does not influence the solution, at least in the range 0.1 - 0.9; a value of 0.5 has been used in the simulation).
The boundary conditions are the usual ones, dictated by the characteristics analysis (the flow is assumed to be sub-critical):
q_θ and q_r are imposed at the entrance, y at the outlet,
q_r is assumed to be zero at the walls (described by the equations $r=r_{min}$, $r=r_{max}$).
For the system of equations (1) and (5) a Preissmann scheme has been used, in its classical form, after having integrated the equations over a step along the θ-direction and in time.
Both equations have the form:

$$\frac{\partial A}{\partial t} + \frac{1}{r}\frac{\partial B}{\partial \theta} + C = 0$$

with:

$$A = q_\theta$$

$$B = \left(\frac{q_\theta^2}{y} + \frac{1}{2} g\, y^2 \right)$$

$$C = \frac{\partial}{\partial r}\left(\frac{q_r q_\theta}{y} \right) + \frac{2}{r}\, \frac{q_r q_\theta}{y} + g\, y\, \frac{1}{r}\frac{\partial z_f}{\partial \theta} + \frac{1}{\rho}\, \tau_{b\theta}$$

for the equation (1), and

$$A = y$$

$$B = q_\theta$$

$$C = \alpha\, \frac{1}{r}\, q_r$$

for the equation (5).
Integration and discretization according with the Preissmann scheme gives:

$$\Delta\, \mathcal{F} = \mathcal{F}^{k+1} - \mathcal{F}^k = \mathcal{G}^k$$

with:

$$\mathcal{F} = \Delta\theta\, [\, \Phi\, A_{i+1,\,j} + (1-\Phi)\, A_{i,\,j}\,] + \Delta t\, \Theta\, \frac{1}{r}\, [\, B_{i+1,\,j} - B_{i,\,j}\,]$$
$$+ \Delta t\, \Delta\theta\, \Theta\, [\, \Phi\, C_{i+1,\,j} + (1-\Phi)\, C_{i,\,j}\,]$$

and

$$\mathcal{G} = -\, \Delta t\, \frac{1}{r}\, [\, B_{i+1,\,j} - B_{i,\,j}\,] - \Delta t\, \Delta\theta\, [\, \Phi\, C_{i+1,\,j} + (1-\Phi)\, C_{i,\,j}\,]$$

where the superscript k indicates that the value is referred to the time $k\cdot\Delta t$, the subscript i that the value is referred to the $i\cdot\Delta\theta$ position along the θ-axis, and the subscript j that the value is referred to the $j\cdot\Delta r$

position along the r-axis.

The symbols Θ and Φ (with values between 0 and 1) indicate the weights (in time and space) of the discretization.

The derivatives appearing in the expression of C for eq. (1) have been discretized as centred differences, excluding the points at the boundaries, where side differences have been used.

The $\mathcal{F}$ and $\mathcal{G}$ are both functions of the state variables $q_{\theta\ i, j}$, $q_{\theta\ i+1, j}$, $y_{i, j}$, $y_{i+1, j}$ (the q_r and z_f being considered known, with their values at the preceding time step), so a linearised expression of $\Delta\mathcal{F}$ can be written in the form:

$$\Delta\mathcal{F} = \frac{\partial\mathcal{F}}{\partial y_{i+1, j}} \Delta y_{i+1, j} + \frac{\partial\mathcal{F}}{\partial q_{\theta\ i+1, j}} \Delta q_{\theta\ i+1, j} + \frac{\partial\mathcal{F}}{\partial y_{i, j}} \Delta y_{i, j}$$

$$+ \frac{\partial\mathcal{F}}{\partial q_{\theta\ i, j}} \Delta q_{\theta\ i, j}$$

and we get for every "row" (of n nodes) a linear system of $2 \cdot (n-1)$ equations of the type:

$$a' \Delta y_{i+1, j} + b' \Delta q_{\theta\ i+1, j} + c' \Delta y_{i, j} + d' \Delta q_{\theta\ i, j} + e' = 0$$

$$a'' \Delta y_{i+1, j} + b'' \Delta q_{\theta\ i+1, j} + c'' \Delta y_{i, j} + d'' \Delta q_{\theta\ i, j} + e'' = 0$$

with an obvious meaning of the symbols ($e = -\mathcal{G}$), where the superscript ' indicates that the equation (1) has been considered, and the superscript " indicates that the equation (5) has been taken into account. The coefficients can be calculated analytically.

The two boundary conditions for $i = i_{min}$ and for $i = i_{max}$ allow the solution for the $2 \cdot n$ state variables q_{θ} and y. After the procedure has been extended to all the m "rows", the state variables are updated.

The linear system is tridiagonal and its solution is trivial. We have adopted the so-called double-sweep technique.

The procedure which has been used for the r-direction is slightly different and originally developed.

The equations (2) and (6) can both be written as:

$$\frac{\partial A}{\partial t} + \frac{1}{r} \frac{\partial B}{\partial \theta} + \frac{\partial C}{\partial r} + D = 0$$

with

$$A = q_r$$

$$B = \left(\frac{q_r q_\theta}{y} \right)$$

$$C = \left(\frac{q_r^2}{y} + \frac{1}{2} g y^2 \right)$$

$$D = \frac{1}{r} \frac{q_r{}^2}{y} - \frac{1}{r} \frac{q_\theta{}^2}{y} + g\,y\,\frac{\partial z_f}{\partial r} + \frac{1}{\rho}\,\tau_{br}$$

for the equation (2), and

$$A = y$$
$$B = 0$$
$$C = q_r$$
$$D = (1-\alpha)\,\frac{1}{r}\,q_r$$

for the equation (6).

The Preissmann discretization is here done again after the integration over a step in space (but in both directions now) and in time, obtaining for any mesh element an expression formally similar to the preceding one:

$$\Delta\,\mathcal{F} = \mathcal{F}^{k+1} - \mathcal{F}^k = \mathcal{G}^k$$

but with rather more complex expressions for $\mathcal{F}$ and $\mathcal{G}$:

$$\mathcal{F} = \Delta\theta\,\Delta r\,\{\,\Phi\,[\,\Psi\,A_{i+1,\,j+1} + (1-\Psi)\,A_{i+1,\,j}\,]$$

$$+ (1-\Phi)\,[\,\Psi\,A_{i,\,j+1} + (1-\Psi)\,A_{i,\,j}\,]\,\}$$

$$+ \Delta t\,\Delta r\,\Theta\,\{\,\Psi\,\frac{1}{r_{j+1}}\,[\,B_{i+1,\,j+1} - B_{i,\,j+1}\,]$$

$$+ (1-\Psi)\,[\,B_{i+1,\,j} - B_{i,\,j}\,]\,\}$$

$$+ \Delta t\,\Delta\theta\,\Theta\,\{\,\Phi\,[\,C_{i+1,\,j+1} - C_{i+1,\,j}\,] + (1-\Phi)\,[\,C_{i,\,j+1} - C_{i,\,j}\,]\,\}$$

$$+ \Delta t\,\Delta\theta\,\Delta r\,\Theta\{\,\Phi\,[\,\Psi\,D_{i+1,\,j+1} + (1-\Psi)\,D_{i+1,\,j}]$$

$$+ (1-\Phi)\,[\,\Psi\,D_{i,\,j+1} + (1-\Psi)\,D_{i,\,j}]\,\}$$

$$\mathcal{G} = -\,\Delta t\,\Delta r\,\{\,\Psi\,\frac{1}{r_{j+1}}\,[\,B_{i+1,\,j+1} - B_{i,\,j+1}\,]$$

$$+ (1-\Psi)\,[\,B_{i+1,\,j} - B_{i,\,j}\,]\,\}$$

$$-\,\Delta t\,\Delta\theta\,\{\,\Phi\,[\,C_{i+1,\,j+1} - C_{i+1,\,j}\,] + (1-\Phi)\,[\,C_{i,\,j+1} - C_{i,\,j}\,]\,\}$$

$$- \Delta t \, \Delta \theta \, \Delta r \, \{ \, \Phi \, [\, \Psi \, D_{i+1, \, j+1} + (1 - \Psi) \, D_{i+1, \, j}]$$

$$+ (1 - \Phi) \, [\, \Psi \, D_{i, \, j+1} + (1 - \Psi) \, D_{i, \, j}] \, \}$$

Here a new weight Ψ (for the r-direction) has been introduced.
The $\mathcal{F}$ and $\mathcal{G}$ are now functions of eight state variables: $q_{r \, i, \, j}$, $q_{r \, i+1, \, j}$, $q_{r \, i, \, j+1}$, $q_{r \, i+1, \, j+1}$, $y_{i, \, j}$, $y_{i+1, \, j}$, $y_{i, \, j+1}$, $y_{i+1, \, j+1}$, and the linearised expression of $\Delta \mathcal{F}$ must now be written as:

$$\Delta \mathcal{F} = \frac{\partial \mathcal{F}}{\partial y_{i+1, \, j}} \Delta y_{i+1, \, j} + \frac{\partial \mathcal{F}}{\partial q_{r \, i+1, \, j}} \Delta q_{r \, i+1, \, j} + \frac{\partial \mathcal{F}}{\partial y_{i, \, j}} \Delta y_{i, \, j}$$

$$+ \frac{\partial \mathcal{F}}{\partial q_{r \, i, \, j}} \Delta q_{r \, i, \, j} + \frac{\partial \mathcal{F}}{\partial y_{i+1, \, j+1}} \Delta y_{i+1, \, j+1} + \frac{\partial \mathcal{F}}{\partial q_{r \, i+1, \, j+1}} \Delta q_{r \, i+1, \, j+1}$$

$$+ \frac{\partial \mathcal{F}}{\partial y_{i, \, j+1}} \Delta y_{i, \, j+1} + \frac{\partial \mathcal{F}}{\partial q_{r \, i, \, j+1}} \Delta q_{r \, i, \, j+1}$$

It should anyway be reminded that the n "columns" can be analysed from upwards to downwards, having for the first one $i = i_{min}$, where the values of Δq_r are known (imposed by the boundary conditions). It suggests to choose as four unknowns the values of $\Delta y_{i, \, j}$, $\Delta y_{i, \, j+1}$, $\Delta q_{r \, i+1, \, j}$ and $\Delta q_{r \, i+1, \, j+1}$, simply assuming to be negligible the contribution of the terms where $\Delta y_{i+1, \, j}$ and $\Delta y_{i+1, \, j+1}$ appear; of course their values will be calculated in the subsequent "column".
A possible choice is the use of an iterative technique, adopting for the second iteration step the values of $\Delta y_{i+1, \, j}$ and $\Delta y_{i+1, \, j+1}$ obtained in the first one, and so on. Some tests have shown that it is not necessary, as the results of the first step do not differ appreciably from the iterated ones.
So we get again a linear system of $2 \cdot (m-1)$ equations, similar to the previous one:

$$a' \, \Delta y_{i, \, j+1} + b' \, \Delta q_{r \, i+1, \, j+1} + c' \, \Delta y_{i, \, j} + d' \, \Delta q_{r \, i+1, \, j} + e' = 0$$

$$a'' \, \Delta y_{i, \, j+1} + b'' \, \Delta q_{r \, i+1, \, j+1} + c'' \, \Delta y_{i, \, j} + d'' \, \Delta q_{r \, i+1, \, j} + e'' = 0$$

with coefficients that can be again analytically calculated (remind that e' and e" include now at least the terms in $\Delta q_{r \, i, \, j}$ and in $\Delta q_{r \, i, \, j+1}$). The boundary conditions for $j = j_{min}$ and for $j = j_{max}$ allow the solution for the $2 \cdot m$ state variables of the "column"; the following "column" is now in the same conditions of the previous one, and the same procedure can be used. For the last one we cannot calculate the final Δy, but they are given by the outlet boundary conditions.

A new updating for y (and of course for q_r) closes the second step.

It should be noticed that when a steady or quasi-steady flow is considered, every time step may be seen as an iteration of a Gauss procedure, and the problem can be considered entirely solved in its non-linear form.

The present technique has been developed and adopted after the checking of several different existing procedures, this one having shown to be more stable and robust.

No comment is necessary for the solution of equation (4), which has been solved explicitly at every time step.

THE CASE CONSIDERED

After some preliminary tests, which have shown the strong influence on the bed evolution of the boundary conditions (in particular of the distribution of the inflowing solid discharge), we have focused our attention on a case similar to the case experimentally studied by Hooke, R. L. (1975), which has been considered also by N. Struiksma, K.W. Olesen, C. Flokstra and H.J. de Vriend (1985). The experiments were conducted in the Meander Flume at the University of Uppsala, where a succession of identical meanders has been realised. The original lay-out presents a sinusoidal plane development, and was designed in such a way that the bed deformation agreed with the observations of Leopold et al. (1964). We have approximated the sinusoidal lay out by a succession of circular bends (opening angle 120°, inner radius 2.7 m, outer radius 3.7 m, mean curvature radius 3.2 m); the width is then of 1.0 m, and a constant discharge of 0.02 m³/s was flowing over a bed of monogranular sediments with a mean diameter of 0.3 mm. The mean water depth was of 0.073 m.

The simulation has been conducted with a grid of 1° for the θ-direction, and 0.05 m for the r-direction. The time step was of 1 s. The resulting Courant number is over 20. A larger Courant number could have been used, but we did not test sufficiently the accuracy of the solution for very high Courant numbers, and we limited ourselves to a rather standard value.

A value of $\Theta = 1$ has been chosen (fully implicit scheme), while both Φ and Ψ have been put equal to 0.5.

For the inlet boundary conditions, the unit-width discharge distribution has been put at every step identical (and symmetrical) to the resulting distribution at the outlet, and similarly has been done for the entering solid discharge. A parallel technique has been followed for the water level distribution of the outlet. In this way a strong influence of upstream and downstream boundary conditions should be excluded, being the considered bend one of a succession (theoretically unlimited) of identical and symmetrical bends. Fig. 1 illustrates the considered geometrical scheme.

For the sediments discharge the formula of Engelund and Hansen has been used (as suggested also by N. Struiksma, K.W. Olesen, C. Flokstra and H.J. de Vriend), but in its general form, without any corrections for the local bed slope or for helical flow effects. This choice is always due to the idea to test the scheme in a first stage in his more elementary and solid form.

RESULTS AND COMPARISON

The simulation has been conducted for a period longer than eleven hours,

and the regime solution was reached in about three ours, remaining unchanged for the following time.
It has immediately become evident that the results of the simulation cannot be compared with the experimental data. As an example we show in fig. 2 and fig. 3 the comparison between the computed and the measured relative water depth (referred to the average depth) at a distance of 15% of the width from the two vertical walls.
Qualitatively and quantitatively the results are so different, that the small geometrical difference between the physical and the theoretical models cannot be invoked.
The resul s of the shallow-water scheme seems to be rather "honest" in general, b.it the bed dynamics is evidently dominated by the three-dimensional flow pattern (mainly the helical flow). Defining a "node" of the solution a section where a physical quantity assumes a constant value over the whole cross-section, we have found that "nodes" occupies different positions for different physical quantities, as shown in fig. 4, 5, 6 and 7, where some general features of the regime solution are shown.

CONCLUDING REMARKS

We have tested the shallow-water scheme for a schematic alluvial river reach, developing a robust technique for solving the flow equations, which have been used in a polar coordinate system.
It is probably too early to conclude that the shallow-water scheme is definitively unsuitable for computing bed evolution in a river reach (the complete research program will be carried out), but the strong influence of factors which are not taken into account by the vertical averaging seems to play a dominant role in the phenomenon.

ACKNOWLEDGEMENT

This research has been supported by a grant of ENEL-CRIS.

REFERENCE

1. Colombini, M. and M. Tubino, 1990, Risonanza non lineare in meandri di ampiezza finita. Parte seconda: risultati sperimentali. XXII Convegno di Idraulica e Costruzioni Idrauliche, Cosenza.

2. Engelund, F., 1976, Experiments in curved alluvial channel, part 2, progress Rep. 38, Tech. Univ. Denmark.

3. Ikeda, S. and T. Nishimura, 1985, Bed topography in bends of sand-silt rivers, J. Hydr. Engrg., ASCE, Vol. 111, No. 11.

4. Ikeda, S. and T. Nishimura, 1986, Flow and bed profile in meandering sand-silt rivers, J. Hydr. Engrg., ASCE, Vol. 112, No. 7.

5. Hooke, R.L., 1975, Distribution of sediment transport and shear stress in a meander bend, Journal of Geology, Vol. 83, No. 5.

6. Kalkwijk, J.P.Th. and H.J. de Vriend, 1980, Computation of the flow in shallow river bends, J. Hydr. Res., IAHR, vol. 18, No. 4.

7. Koch, F.G. and C. Flokstra, 1980, Bed level computations for curved alluvial channels. Proc. XIX IAHR Congr., New Delhi, India, vol. 2.

8. Leopold, L.B., Wolman, M.G. and J.P. Miller, 1964, *Fluvial processes in geomorphology*, Freeman and Co., San Francisco.

9. Odgaard, A.J., 1981, Transverse bed slope in alluvial channel bends, J. Hydr. Div. ASCE, vol. 107, No. HY12.

10. Olesen, K.W., 1985, A mathematical model of the flow and bed topography in curved channels, Delft Univ. of Technol., Dept. Civil Engrg., Rep. 85-1.

11. Rozovskii, I.L., 1961, Flow of water in bends of open channels. Israel Progr. for Scientific Transl., Jerusalem.

12. Struiksma, N., Olesen, K.W., Flokstra, C. and H.J. de Vriend, 1985, Bed deformation in alluvial channel bends, J. Hydr. Res., IAHR, vol. 23, No. 1.

13. Talmon, A.M., 1992, Bed topography of river bends with suspended sediment transport. Techn. Univ. Delft, Communications on Hydraulic and Geotechnical Engineering, No. 92-5.

14. Tubino, M., 1989, Linear and nonlinear theory of river meanders. Excerpta, vol. 4.

15. Zimmermann C. and J.F. Kennedy, 1978, Transverse bed slopes in curved alluvial streams, J. Hydr. Div. ASCE, vol. 104, No. HY1.

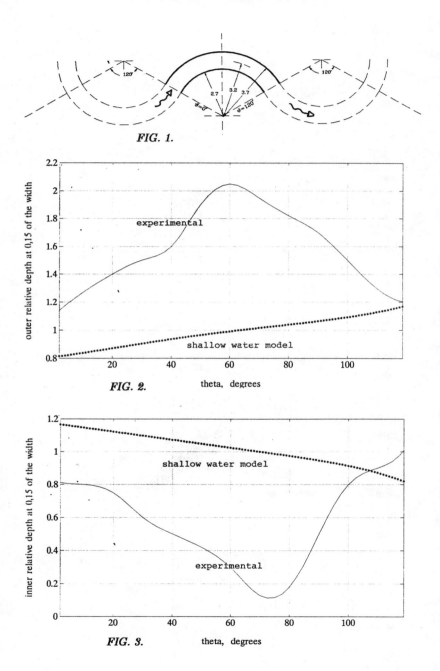

FIG. 1.

FIG. 2.   theta, degrees

FIG. 3.   theta, degrees

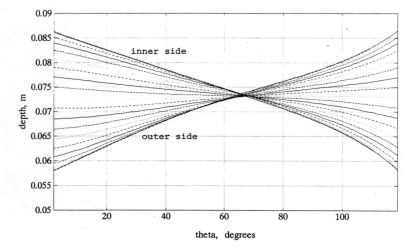

FIG. 4.

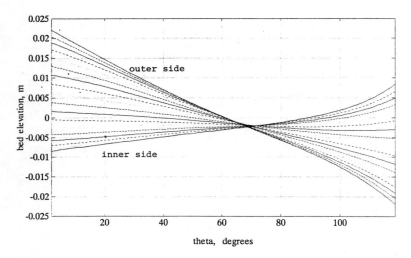

FIG. 5.

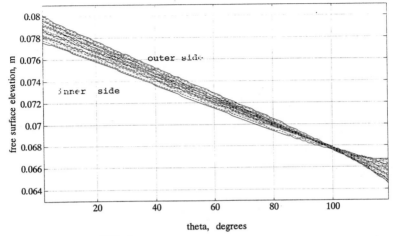

FIG. 6.

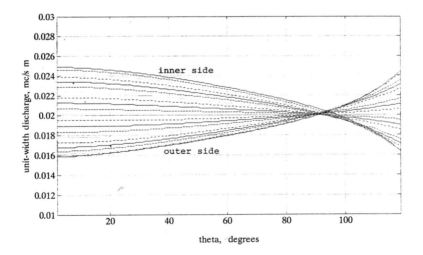

FIG. 7.

Subject Index
Page number refers to first page of paper

Author Index
Page number refers to the first page of paper